Xingni Zhou, Qiguang Miao and Lei Feng
Programming in C

Also of interest

Programming in C, vol. 1: Basic Data Structures and Program Statements
Xingni Zhou, Qiguang Miao, Lei Feng, 2020
ISBN 978-3-11-069117-7, e-ISBN (PDF) 978-3-11-069232-7,
e-ISBN (EPUB) 978-3-11-069249-5

C++ Programming
Li Zheng, Yuan Dong, Fang Yang, 2019
ISBN 978-3-11-046943-1, e-ISBN (PDF) 978-3-11-047197-7,
e-ISBN (EPUB) 978-3-11-047066-6

Elementary Synchronous Programming
Ali S. Janfada, 2019
ISBN 978-3-11-061549-4, e-ISBN (PDF) 978-3-11-061648-4,
e-ISBN (EPUB) 978-3-11-061673-6

MATLAB® Programming
Dingyü Xue, 2020
ISBN 978-3-11-066356-3, e-ISBN (PDF) 978-3-11-066695-3,
e-ISBN (EPUB) 978-3-11-066370-9

Programming in C++
Laxmisha Rai, 2019
ISBN 978-3-11-059539-0, e-ISBN (PDF) 978-3-11-059384-6,
e-ISBN (EPUB) 978-3-11-059295-5

Xingni Zhou, Qiguang Miao and Lei Feng

Programming in C

Volume 2: Composite Data Structures
and Modularization

DE GRUYTER

Authors

Prof. Xingni Zhou
School of Telecommunication Engineering
Xidian University
Xi'an, Shaanxi Province
People's Republic of China
xnzhou@xidian.edu.cn

Qiguang Miao
School of Computer Science
Xidian University
Xi'an, Shaanxi Province
People's Republic of China
qgmiao@xidian.edu.cn

Lei Feng
School of Telecommunication Engineering
Xidian University
Xi'an, Shaanxi Province
People's Republic of China
fenglei@mail.xidian.edu.cn

ISBN 978-3-11-069229-7
e-ISBN (PDF) 978-3-11-069230-3
e-ISBN (EPUB) 978-3-11-069250-1

Library of Congress Control Number: 2020941966

Bibliographic information published by the Deutsche Nationalbibliothek
The Deutsche Nationalbibliothek lists this publication in the Deutsche Nationalbibliografie;
detailed bibliographic data are available on the Internet at http://dnb.dnb.de.

© 2020 Walter de Gruyter GmbH, Berlin/Boston
Cover image: RomoloTavani/iStock/Getty Images Plus
Typesetting: Integra Software Services Pvt. Ltd.
Printing and binding: CPI books GmbH, Leck

www.degruyter.com

Contents

1 Arrays

Main contents
- Concept, usage, and available methods of arrays
- Introduction of representation and nature of arrays through comparison between array/ array elements and plain variables
- Storage characteristics and debugging techniques of arrays
- Programming techniques of multidimensional arrays
- Top-down algorithm design practices

Learning objectives   **!**
- Know how to define and initialize arrays as well as how to access array elements
- Be able to define and use multidimensional arrays
- Know how to deal with character arrays

1.1 Concept of arrays

Program statements and data construct programs. They are sequences of instructions created through algorithm design that conform to program control structures. However, are we able to solve all problems after learning statements, basic data types, program control structures, and algorithm implementation methods of C?

Let us look at a few problems in practice.

1.1.1 Processing data of the same type

Case study 1 Cracking Caesar code

Mr. Brown received an email from his son Daniel. However, the contents seemed a little weird for an email sent by someone in elementary school: it was a meaningless sequence "lettc fmvxlhec hehhc pszi csy".

It later turned out that Daniel read a story of Julius Caesar and created an encrypted email using Caesar code to see whether his father could decrypt it.

During Roman times, Caesar invented the Caesar code to protect the information he exchanged with his generals on the front line from being intercepted by enemy spies. Encryption and decryption of Caesar code were done by shifting letters by a fixed number of positions. The plaintext alphabet was shifted forward or backward by a fixed number of positions to create the ciphertext alphabet. The number of positions shifted was the key for encryption and decryption of Caesar code, as shown in Figure 1.1.

https://doi.org/10.1515/9783110692303-001

Case study 1

Encryption and decryption of Caesar code

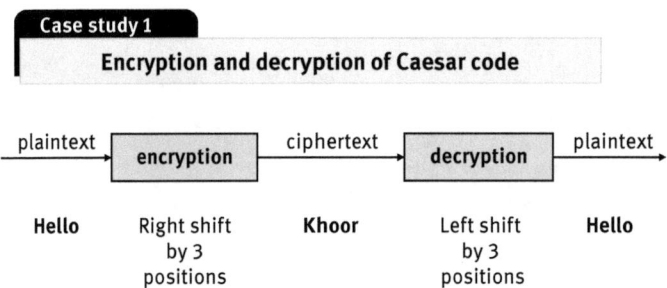

Figure 1.1: Encryption and decryption of Caesar code.

Mr. Brown stared at the ciphertext and thought that it would not be hard to design an algorithm to solve the problem. He could simply shift each character in the ciphertext by one position in the alphabet and repeat this process 26 times to list all possible plaintexts, in which the one that is not nonsense would be the real plaintext. A universal algorithm could be designed using this technique to crack ciphertexts of arbitrary length.

- If the length of the ciphertext is 2, we shift letters by one position in the alphabet each time and list all 26 possible plaintexts.
- If the length of the ciphertext is 10, we shift letters by one position in the alphabet each time and list all 26 possible plaintexts.
- If the length of the ciphertext is 100, we shift letters by one position in the alphabet each time and list all 26 possible plaintexts.

!

Think and discuss Necessary variables in password cracking
1. How many variables are necessary for a program to handle 100 characters?
2. How should we use these variables so that the program handles data in a convenient and unified manner?

Discussion: Solving a problem with computers involves two major steps: first, we should use reasonable data structures to describe the problem to store data into computers; second, we create algorithms to solve it. To answer the earlier questions, we need to find a mechanism that describes variables of the same type and handles them consistently.

Code implementation of the algorithm that solves Caesar codes is rather complicated, so we shall introduce it later. Before that, let us consider a reversed order problem that is more trivial.

Case study 2 Reversing 100 numbers

Write a program that reads 100 numbers and outputs them in a reversed order.

We are going to focus on how to handle variables of the same type. For a simpler description, we use variables with subscripts to represent the numbers, as shown in Figure 1.2.

Case study 2

Reversing 100 numbers

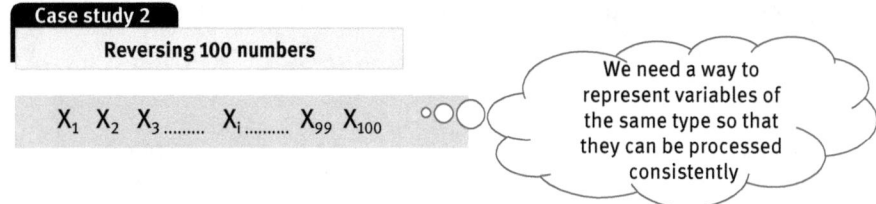

Figure 1.2: Representation of 100 variables of the same type.

The flow of outputting 100 numbers backward is given in Figure 1.3. The program reads the numbers in a loop starting from X_1, and outputs them using a loop starting from X_{100}.

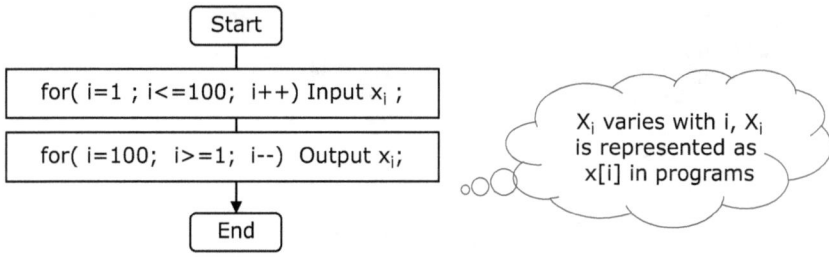

Figure 1.3: Flow of outputting numbers backward.

Variable X_i is uniquely identified by the value of i. We use x[i] to represent X_i in programs so we can type the names using keyboards.

The code implementation is as follows:

```
01 int main(void )
02 {
03    int i;
04    int x[100];
05    for ( i=1; i<=100; i++) scanf ("%d", &x[i] );
06    for ( i=100; i>=1; i-- ) printf ("%d", x[i] );
07    return 0;
08 }
```

On line 4, the statement defines 100 variables with subscripts of type int. It is more convenient to "batch" define variables of the same type.

It is worth noting that the starting subscripts on line 5 and line 6 do not follow the convention of using C arrays exactly.

Subscripts of arrays start from 0 in C. Here, we are trying to make the flow more intuitive by not following this rule.

Case study 3 Simple table processing
Write a program that calculates the average grade of a student in six courses.

Figure 1.4 shows how to store grades and pseudo code of the algorithm. To store grades, we use six variables, namely grade 0 to grade 5. On line 1 of the pseudo code, int grade[6] defines 6 int variables. Note that the number of variables is 6, but the range of subscripts is 0 to 5.

Case study 3

Simple table processing

Course 1	Course 2	Course 3	Course 4	Course 5	Course 6	Average
80	82	91	68	77	78	

i	0	1	2	3	4	5
grade[i]	80	82	91	68	77	78

Pseudo code
Store grades in int grade[6]
Total score total = 0; Counter i = 0;
while i < 6
total= total+grade[i];
i++;
Average= total / 6

Figure 1.4: Simple table processing.

We use a while loop to add each grade[i] to total grade total. The value of i increases in each iteration so that all variables are handled.

It is clear that the algorithm is trivial as long as we find a way to store and represent data of the same type. This also shows that the way data are organized and represented is a crucial issue when solving problems with computers.

Case study 4 Complex table processing
Suppose there are four students, all of which take the same six courses. Write a program that calculates average grades for each of them.

The only difference between this problem and the one earlier is the number of grades. As shown in Figure 1.5, we can use a two-dimensional table to store data. Its row index and column index uniquely identify a grade. For example, the grade in row 1 column 2, whose value is 82, can be represented by grade[1][2].

Case study 4

Complex table processing

ID	Course 1	Course 2	Course 3	Course 4	Course 5	Course 6	Average
1001	80	82	91	68	77	78	
1002	78	83	82	72	80	66	
1003	73	50	62	60	75	72	
1004	82	87	89	79	81	92	

Column j

Grade of student 1 in course 2 is grade[1][2]=82

grade[i][j]	j=0	j=1	j=2	j=3	j=4	j=5
i=0	80	82	91	68	77	78
i=1	78	83	82	72	80	66
i=2	73	58	62	60	75	72
i=3	82	87	89	79	81	92

Row i

Figure 1.5: Complex table processing.

We can use a for loop to process grades for a single student and use another one to calculate average grades for all of them. The algorithm and code implementation will be given in the section of two-dimensional arrays.

1.1.2 Representation of data of the same type

The discussion earlier showed that a new mechanism is necessary to handle data of the same type. With respect to data representation and processing, arrays are a data structure that regularly expresses data so that they are processed regularly.

Since arrays are collections of variables whose names have a pattern, they are supposed to have features of variables. Figure 1.6 compares arrays with plain variables.

		Plain variable	Array	Notes
Definition		type name;	To be determined, but it should consists of: type, name, number of variables	• Memory is allocated upon definition • Size of memory allocated is determined by variable type
Storage unit	Quantity	One	Multiple	Each storage unit of an array has the same size
	Length	sizeof(type)	sizeof(type)* number of variables	Length is measured in bytes
	Address	&name	To be determined	
Referencing method		name	name[index]	
Initialization		type name=value	To be determined	It is easier to process in programs if variables are initialized

Figure 1.6: Comparison of a group of variables with a single variable.

During the definition of a plain variable, the system allocates memory according to its type specified by programmers. The definition of an array consists of type, name and, in particular, the number of variables in the array.

There are multiple variable values in an array, so they should be stored in multiple storage units, whose sizes depend on types of the variables. The size of a storage unit is measured in bytes and can be computed using the sizeof operator.

Besides, a referencing method of the address of a storage unit is necessary so that programmers can inspect the unit.

We can infer from the examples earlier that the referencing method of variable values in an array is to use the array name with an index.

Moreover, we should be able to initialize an array since we can do the same with plain variables. Hence, a corresponding syntax is necessary.

1.2 Storage of arrays

There are four issues related to array storage, namely definition, initialization, memory allocation, and memory inspection.

1.2.1 Definition of arrays

1.2.1.1 Definition of arrays

An array is a collection of data of the same type. Figure 1.7 shows how to define an array, where a definition is constructed by a type identifier followed by an array name and multiple constants inside square brackets. Each constant indicates the number of variables in the corresponding dimension.

Arrays

An array is a collection of data of the same type.

Memory is allocated upon definition, which remains unchanged during execution

Syntax

type name [constant 1][constant 2] ......[constant n];

E.g.

Definition	Type	Name	Number of dimensions	Number of variables	Memory size
int x[100]	int	x	1	100	100* sizeof(int)
char c[2][3]	char	c	2	2*3	2*3* sizeof(char)

Figure 1.7: Definition of arrays.

In the figure above, the first row defines a one-dimensional integer array x with 100 variables. To compute the size of its memory space, we can obtain the size of its type using the sizeof operator and multiply it with the number of variables. The second row defines a two-dimensional character array with two rows and three columns. In other words, it has six variables in total. The array name is c.

1.2.1.2 Reference of array elements

C uses a special term for variables in an array: array elements. An array element is used in the same way as a single variable. To reference an array element, we use the array name suffixed by an index wrapped in square brackets.

Think and discuss Do contents inside square brackets in an array definition and an element reference refer to the same thing?

Discussion: The index of an array element is a numerical expression, which indicates the position of the element in an array; the object inside square brackets in an array definition has to be a constant, which indicates the number of elements in the corresponding dimension. It is worth noting that the number of elements must not be a variable. Like plain variables, arrays obtain memory space from the system during array definition. The size of the allocated space does not change during execution once the array is defined. Such a way of memory utilization and management is called static memory allocation. On the other hand, C also provides "dynamic memory allocation," which will be introduced in examples in chapter "Functions".

Indices of array elements in C must start from 0. Accessing an array out of bound leads to a logic error, but it is not a syntax error.

For example, the one-dimensional array x defined in Figure 1.8 has 100 elements with an index range 0 to 99. If we try to access an element outside this range, we are accessing the array out of bound. Grammatically, it is equivalent to using undefined variables.

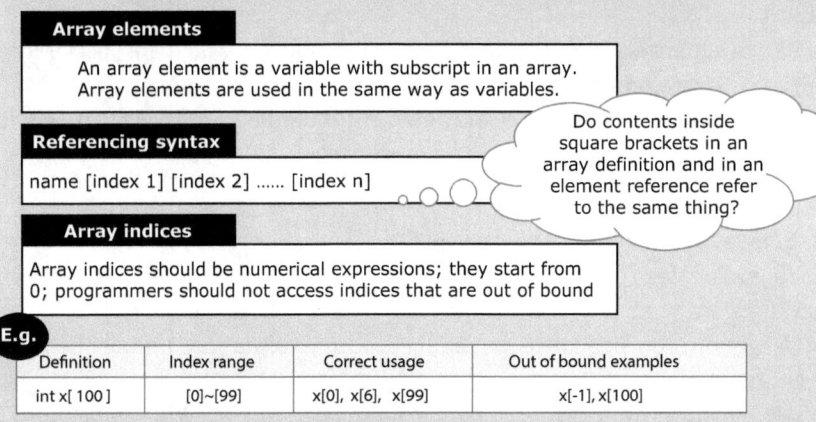

Array elements
An array element is a variable with subscript in an array.
Array elements are used in the same way as variables.

Do contents inside square brackets in an array definition and in an element reference refer to the same thing?

Referencing syntax
name [index 1] [index 2] ...... [index n]

Array indices
Array indices should be numerical expressions; they start from 0; programmers should not access indices that are out of bound

E.g.

Definition	Index range	Correct usage	Out of bound examples
int x[100]	[0]~[99]	x[0], x[6], x[99]	x[-1], x[100]

Figure 1.8: Array elements and their referencing rules.

The reason that out-of-bound errors are not syntax errors is that the compiler will not check whether the index is valid. As a result, programmers should take care of indices when using arrays.

Knowledge ABC Index out-of-bound errors
An index out-of-bound error happens when accessing an array element whose index does not fall in the predefined index range. C compilers seldom check whether indices are valid. Accessing an index that is out of bound leads to the following issues.

First, although reading an out-of-bound element does not change values in memory, the calculation that uses this value will generate wrong results.

Second, writing to an out-of-bound element does change values in memory. If the memory units we write to contain values of other variables, the program may also generate wrong results. Furthermore, it is tough to debug in this case, since we do not know when the modified value gets referenced.

An index out-of-bound error may occur in arrays and pointers that point to arrays. It is one of the most common mistakes that beginners may make, so we should be careful when using arrays.

Having learned how to define arrays and how to reference array elements, we can complete the program for number reversing problem.

```
01  int main(void )
02  {
03    int i;
04    int x[100];        // Array definition
      //x[i] references array elements, the index is an expression
05    for ( i=0; i<100; i++) scanf ("%d", &x[i] );
06    for ( i=99; i>=0; i-- ) printf ("%d", x[i] );
07    return 0;
08  }
```

Line 4 contains definition of an array. Note how we reference array elements on line 5 and line 6.

Indices in square brackets on line 5 and 6 are variables, which are special forms of expression. They start from 0 and end at 99.

Grammatically, the index of an array element should be a numerical expression and the index of the first element of an array must be 0.

1.2.1.3 Storage characteristics of arrays

The system allocates contiguous memory space to an array based on its definition, so the storage characteristics can be summarized as "memory is allocated during definition, the size keeps unchanged during execution, and elements are stored continuously".

Figure 1.9 shows an array definition written by a student. Will the memory be allocated to array a in this case?

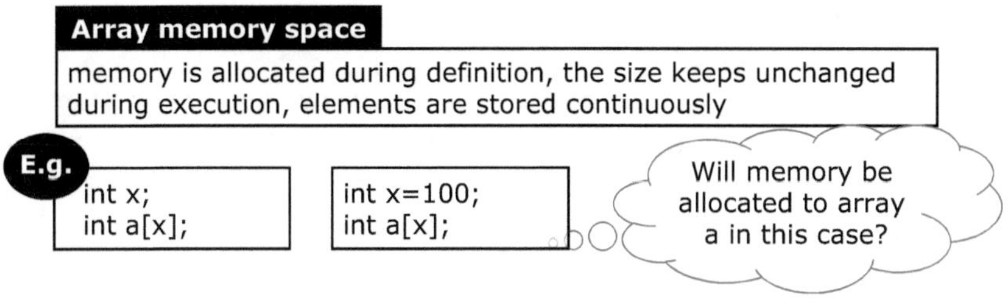

Array memory space

memory is allocated during definition, the size keeps unchanged during execution, elements are stored continuously

E.g.

```
int x;
int a[x];
```

```
int x=100;
int a[x];
```

Will memory be allocated to array a in this case?

Figure 1.9: Storage characteristics of arrays.

1.2.1.4 Comparison of variables of the same type with plain variables

With the rules of arrays in C, we can update the table in Figure 1.6 and obtain Figure 1.10.

		Plain variable	Array	Notes
1	Definition	type name;	type name [constant]...[constant]	Number of dimensions of an array is equal to number of indices
2	Name	Variable name	Array name	Identifiers
3	Variable	One	A group	Array elements are of the same type
4	Storage unit — Quantity	One	Multiple	Elements in an array are stored consecutively
	Storage unit — Length	sizeof(type)	sizeof(type) * number of variables	
	Storage unit — Address	&name	name	Allocated by the system
5	Referencing method	name	name[index]...[index]	Number of dimensions of an array is equal to number of indices
6	Initialization	type name=value	type name[constant] ...[constant] = { a group of initial values}	

Figure 1.10: Comparison of arrays with plain variables.

(1) Number of dimensions of an array is determined by the number of indices, that is, the number of pairs of square brackets. The constant in square brackets indicates the number of elements in an array.
(2) Array names are identifiers.
(3) Values of array elements are of the same type.
(4) When allocating memory space for an array, C allocates a continuous space for all elements and defines that the array name refers to the beginning address of the memory allocated. In other words, array names are addresses.
(5) Array elements are accessed by array name with index.
(6) Initialization is done during definition. The syntax of initialization requires curly brackets.

1.2.2 Initialization of arrays

We can modify the keyboard input part in the code implementation of the number reversing problem so that the array is initialized with values. The revised program is as follows:

```
01 int main(void )
02 {
03   int i; //Defines an array and initializes array elements
04   int x[10]={1,2,3,4,5,6,7,8,9,10};
05   //for ( i=0; i<10; i++) scanf ("%d", &x[i]);
06   for ( i=9; i>=0; i-- ) printf ("%d", x[i]);
07   return 0;
08 }
```

Statement on line 4 defines the array and initializes array elements, so the keyboard input assignment can be skipped.

What is the advantage of initializing an array? If we have to debug the program multiple times, it is more efficient to initialize the array than typing in numbers repeatedly.

Array initialization defines an array and initializes its elements at the same time. There are three ways to initialize an array in C, as shown in Figure 1.11.

Array initialization

An array initialization defines an array and initializes its elements at the same time

E.g.	Case	Example	Array size	Notes
1	Initialize all elements	int m[5]= {1,3,5,7,9}	5	
		int a[2][3] = { {1,3,5}, {2,4,6}};	2 by 3	A 2-d array is stored in a row-first manner
2	Initialize some elements	int b[5] = {1,3,5}	5	Uninitialized elements are set to 0 automatically by the system
		int x[100] ={ 1,3, 5, 7 };	100	
3	Array size determined by number of initial values	int n[ ] = {1,3,5,7,9}	5	
		char c[ ] ="abcde";	6	String termination mark '\0' is also an element

Figure 1.11: Array initialization.

1.2.2.1 Initialize all elements

In the first case in Figure 1.11, the one-dimensional array m has five elements and five values are assigned to the array. The two-dimensional array a has two rows and three columns, so it consists of six elements. Note that how curly brackets are used when assigning all six values.

1.2.2.2 Initialize some elements
In the second case, the length of array b is 5, but only the first three elements are initialized with a value. The other elements are automatically initialized with 0 by the C language system.

1.2.2.3 Array size determined by number of initial values
We can omit the array size in square brackets when defining arrays. The size can be determined by the system based on the number of initial values. In particular, C allows us to assign initial values to character arrays with strings. Note that the string termination mark '\0' is an element as well.

1.2.3 Memory layout of arrays

We will introduce the memory layout of arrays through examples.

1.2.3.1 Memory layout of one-dimensional arrays
A one-dimensional array x of size 100 is defined in Figure 1.12. Indices start from 0 and end at 99. The first four elements are initialized with initial values, while the rest are 0. These elements are stored contiguously in the order of index, that is, from x[0] to x[99].

```
int x[ 100 ]={ 1, 3, 5, 7 };
```

C defines that array elements are stored consecutively in the order of indices

Index	0	1	2	3	4	...	i	...	98	99
Element value	1	3	5	7	0	0	0	0	0	0
Element storage order	x[0]	x[1]	x[2]	x[3]	x[4]		x[i]		x[98]	x[99]

Figure 1.12: Memory layout of one-dimensional array.

1.2.3.2 Memory layout of two-dimensional arrays
As shown in Figure 1.13, two-dimensional array a has two rows and three columns. Its elements are stored in a row-first manner.

The 0th row is initialized with 1, 3, and 5, while the first row is initialized with 2, 4, and 6. The 0th row is stored first, followed by the first row. Note that a[0] denotes

```
int a[2][3]={ {1,3,5}, {2,4,6} };
```

```
C defines that the 1 - dimensional form of a 2 - dimensional array denotes "row address"
```

	0	1	2
a[0] → 0	1	3	5
a[1] → 1	2	4	6

Row address	a[0]			a[1]		
Element storage order	a[0][0]	a[0][1]	a[0][2]	a[1][0]	a[1][1]	a[1][2]
Element value	1	3	5	2	4	6

Figure 1.13: Memory layout of two-dimensional array.

the beginning position of the 0th row and a[1] denotes the beginning position of the first row.

C defines that the one-dimensional form of a two-dimensional array which denotes "row address".

1.2.4 Memory inspection of arrays

With the help of IDE, we can inspect how arrays are stored in the memory. We shall start from cases where arrays are initialized. The program is as follows:

```
01 //Use an initial value list to initialize arrays
02 #include <stdio.h>
03 int main(void)
04 {
05   // Use an initial value list to initialize arrays
06   int m[5]={1,3,5,7,9};
07   int n[ ]={2,4,6,8};
08   int x[8]={1,3,5,7};
09   char c[ ]="abcde";
10   int a[2][3]={ {1,3,5}, {2,4,6}};
11   int i, j;
12
13   //Output 1-dimensional array m as a list
14   printf( "1-dimensional array m[5]\n");
15   printf( "%s%13s\n", "Element", "Value" );
16   for ( i = 0; i < 5; i++ )
17   {
18     printf( "%6d%13d\n", i, m[ i ] );
19   }
20   printf( "\n");
21
```

```
22  // Output 2-dimensional array m as a list
23  printf( "2-dimensional array a[2][3]\n");
24  for (i = 0; i < 2; i++) //Row index range
25  {
26    for (j = 0; j < 3; j++) //Column index range
27    {
28      printf( "%d ", a[i][j] );
29    }
30    printf( "\n");
31  }
32  return 0;
33 }
```

On line 6, we define an integer array m of size 5 and initialize it. If we type in the array name m in the Watch window, we can see the beginning address of the array and values of each element, as shown in Figure 1.14.

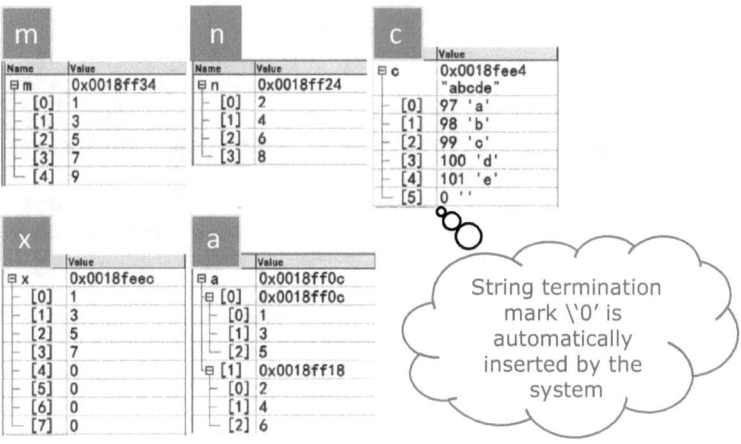

Figure 1.14: Inspecting memory of array 1.

On line 7, we define an integer array n without specifying the size and initialize it with four initial values. We can see that 4 memory units are allocated to it.

On line 8, we define an integer array x of size 8 and partially initialize it. It is clear that the uninitialized elements are set to 0 by the system.

On line 9, we define a character array c without specifying the size and initialize it with a string of five characters. The system allocates six storage units, where the last one has value 0. This is the string termination mark inserted by the system automatically. It also takes up one storage unit.

On line 10, we define a 2 by 3 two-dimensional array a and initialize it. Each row of the array has a beginning address, where the address of the first row is also the beginning address of the entire array.

On line 15, the table header is printed.

On lines 16–19, we use a for loop to output index i and corresponding array elements m[i].

Program result:

```
1-dimensional array m[5]
Element          Value
0                1
1                3
2                5
3                7
4                9
2-dimensional array a[2][3]
1   3   5
2   4   6
```

> **Knowledge ABC** Differences between '\0', '0', "0" and 0 in C
>
> Characters are stored as their ASCII values in C. Each character takes up 1 byte. The first value in the ASCII table is 0, which corresponds to character Null, namely '\0'. It is used as the termination mark of strings and is inserted to the end of strings automatically by the system.
>
> Character '0' has ASCII value 48 or 0 × 30 in hexadecimal form. To convert a number into the corresponding character in a program, for example, converting number 8 into character 8, we can write 8+'0' in the code.
>
> The character '0' is a character literal, while number 0 is an integer literal. They are different objects and are stored in different ways in computers. Character literals can be used as integers in computation.
>
> The difference between "0" and '0' is that "0" is a string literal while '0' is a character literal. They are completely different. Character literals are wrapped by single quotation marks while string literals use double quotation marks. A character literal has to be a single character, while a string literal can have more than *one* character.

The output of two-dimensional array a is implemented by two for loops.

On line 24, the first for loop iterates through row index i, which has range 0 to 1.

On line 26, the second for loop iterates through column index j, which has range 0 to 2.

In Figure 1.15, we can see that j traverses the range 0 to 2 when i is 0, and traverses the range again when i is 1.

	0	1	2
a[0] → 0	1	3	5
a[1] → 1	2	4	6

Row i	0			1		
Column j	0	1	2	0	1	2
a[i][j]	1	3	5	2	4	6

Figure 1.15: Inspecting memory of array 2.

When defining an array, the system allocates contiguous memory space to store its elements based on the array type and number of elements. It is shown in the Memory window that int n[4] takes up a continuous block of memory with size 4*4 bytes (in a 64-bit compiling environment, type int takes up 4 bytes, which can be verified by subtracting addresses of two array elements), as illustrated in Figure 1.16.

Memory		Address	Value	Variable
Address: 0x18ff24		18FF24	2	n[0]
0018FF24 02 00 00 00		18FF28	4	n[1]
0018FF28 04 00 00 00		18FF2C	6	n[2]
0018FF2C 06 00 00 00				
0018FF30 08 00 00 00		18FF30	8	n[3]

Figure 1.16: Continuous storage of a one-dimensional array.

Similarly, int a[2][3] takes up a continuous block of memory with size 6*4 bytes, as shown in Figure 1.17.

Memory	Row	Address	Value	Variable	Row address
Address: 0x18ff0c		18FF0C	1	a[0][0]	
0018FF0C 01 00 00 00	Row 0	18FF10	3	a[0][1]	a[0] 18FF0C
0018FF10 03 00 00 00		18FF14	5	a[0][2]	
0018FF14 05 00 00 00		18FF18	2	a[1][0]	
0018FF18 02 00 00 00	Row 1	18FF1C	4	a[1][1]	a[1] 18FF18
0018FF1C 04 00 00 00		18FF20	6	a[1][2]	
0018FF20 06 00 00 00					

Figure 1.17: Continuous storage of a two-dimensional array.

Note that the array name refers to the address of the entire array, which is also the beginning address of the array.

With rules of storage and elements referencing, we may now process data in arrays.

1.3 Operations on one-dimensional arrays

Example 1.1 Highest score problem

1. Problem description

In the scoring problem we have seen before, there was a step where the highest score was discarded. This is equivalent to finding the maximum of a series of numbers.

2. Algorithm description

We have seen this problem in section "representation of algorithms", where the scores were read from keyboard input. Now we can store scores given by referees in an array score[10]. The algorithm can then be updated accordingly, as shown in Figure 1.18.

Top-level pseudo code	First refinement	Second refinement
Find the highest one of scores stored in array score[10]	Use score[0] as Largest	Counter i=0;
		Largest=score[0];

	Compare each element in array score with Largest, Store the larger in Largest;	while counter i< 10;
		    if(Largest < score[i])
		        Largest=score[i];
		    i increases by 1;

	Output Largest	Output Largest;

Figure 1.18: Eliminating the highest score using an array.

In the second refinement, a counter i is used to record the number of comparisons. Variable Largest is initialized with score[0]; then, Largest is compared with score[i] repeatedly and updated with the larger value in the loop body. Once the loop is done, Largest is printed.

3. Code implementation

```
01 //Finding the maximum number in an array
02 #include <stdio.h>
03 #define SIZE 10
04
05 int main(void)
06 {
07   int score[SIZE]
08       = {89,92,97,95,90,96,94,92,90,98};
09   int i;                        //Counter
10   int Largest =score[0];        //Initialize Largest with score [0] as a comparison basis
11   for ( i = 0; i < SIZE; i++ )
12   {
13     if (Largest < score[i])
14         Largest=score[i];      //Find the maximum
15   }
16   printf( "The highest score is %d\n", Largest );
17   return 0;
18 }
```

Program result:
The highest score is 98

Note: the score array is initialized on line 8 so that testing becomes easier.
On lines 11–15, the for loop finds the largest value and stores it in variable Largest.
Based on this program, it is trivial to write a program that finds the minimum number. Now we can discard both the highest score and the lowest score by replacing them with 0.

4. Debugging
One should carefully design test cases for inspection or verification. Critical points in the debugging of the earlier program are shown in Figure 1.19.

```
Debugging
  plan

- Inspect memory layout of 1-d array
- Reference of array elements
- Use breakpoints to find required values quickly
```

```
11   for ( i= 0; i< SIZE; i++ )
12   {
13       if (Largest < score[i])
14●          Largest=score[i];
             // Find the maximum
15   }
```

Figure 1.19: Debugging the "eliminating highest score" program.

Figure 1.20 shows the score array in the Watch window. There are 10 elements, each of which are initialized with an initial value. The maximum value Largest is initialized with the value of score[0], which is 89.

```c
#include <stdio.h>
#define SIZE 10

int main(void)
{
    int score[SIZE]
    = {89, 92, 97, 95, 90, 96, 94, 92, 90, 98} ;
    int i;
    int Largest =score[0];
    for ( i = 0; i < SIZE; i++ )
    {
        if (Largest < score[i])
            Largest=score[i];
    }

    return 0;
```

Watch	
Name	Value
⊟ score	0x0018ff20
[0]	89
[1]	92
[2]	97
[3]	95
[4]	90
[5]	96
[6]	94
[7]	92
[8]	90
[9]	98
Largest	89

Figure 1.20: Memory inspection of a one-dimensional array 1.

In Figure 1.21, the condition of if statement in the for loop evaluates to false when i = 0, so Largest keeps unchanged.

```
#include <stdio.h>
#define SIZE 10

int main(void)
{
    int score[SIZE]
    = {89, 92, 97, 95, 90, 96, 94, 92, 90, 98};
    int i;
    int Largest =score[0];
    for ( i = 0; i < SIZE; i++ )
    {
        if (Largest < score[i])
            Largest=score[i];
    }

    return 0;
}
```

Watch	
Name	Value
⊟ score	0x0018ff20
[0]	89
[1]	92
[2]	97
[3]	95
[4]	90
[5]	96
[6]	94
[7]	92
[8]	90
[9]	98
Largest	89
i	0
score[i]	89

Figure 1.21: Memory inspection of a one-dimensional array 2.

In Figure 1.22, i becomes 1 after increment and score[1] = 92.

```
#include <stdio.h>
#define SIZE 10

int main(void)
{
    int score[SIZE]
    = {89, 92, 97, 95, 90, 96, 94, 92, 90, 98};
    int i;
    int Largest =score[0];
    for ( i = 0; i < SIZE; i++ )
    {
        if (Largest < score[i])
            Largest=score[i];
    }

    return 0;
}
```

Watch	
Name	Value
⊟ score	0x0018ff20
[0]	89
[1]	92
[2]	97
[3]	95
[4]	90
[5]	96
[6]	94
[7]	92
[8]	90
[9]	98
Largest	89
i	1
score[i]	92

Figure 1.22: Memory inspection of a one-dimensional array 3.

In Figure 1.23, Largest becomes 92 when i = 1.

```
#include <stdio.h>
#define SIZE 10

int main(void)
{
    int score[SIZE]
    = {89, 92, 97, 95, 90, 96, 94, 92, 90, 98} ;
    int i;
    int Largest =score[0];
    for ( i = 0; i < SIZE; i++ )
    {
        if (Largest < score[i])
            Largest=score[i];
    }

    return 0;
}
```

Watch		
Name		Value
⊟ score		0x0018ff20
	[0]	89
	[1]	92
	[2]	97
	[3]	95
	[4]	90
	[5]	96
	[6]	94
	[7]	92
	[8]	90
	[9]	98
Largest		89
i		1
score[i]		92

Figure 1.23: Memory inspection of a one-dimensional array 4.

In Figure 1.24, we insert a breakpoint in the line pointed by the yellow arrow to inspect program execution conveniently. Using the Go command, we can interrupt the program at this statement whenever the condition of if statement evaluates to true. Here i = 2 and score[2] has value 97, which is larger than the value of Largest, 92.

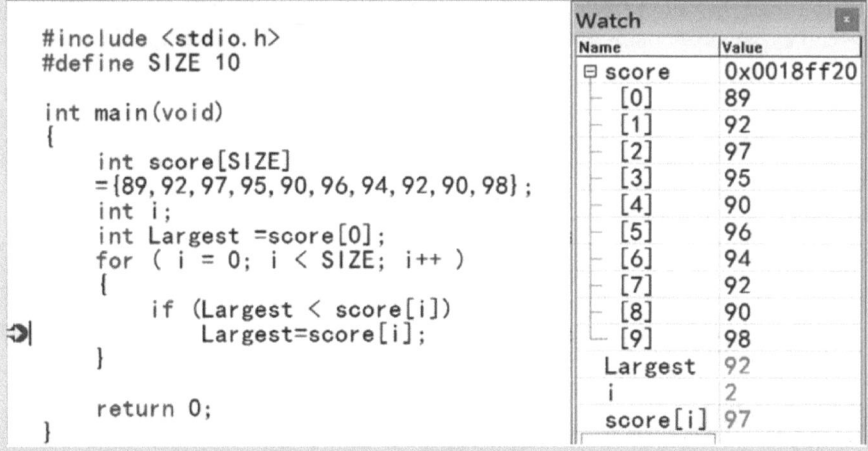

```
#include <stdio.h>
#define SIZE 10

int main(void)
{
    int score[SIZE]
    = {89, 92, 97, 95, 90, 96, 94, 92, 90, 98} ;
    int i;
    int Largest =score[0];
    for ( i = 0; i < SIZE; i++ )
    {
        if (Largest < score[i])
            Largest=score[i];
    }

    return 0;
}
```

Watch		
Name		Value
⊟ score		0x0018ff20
	[0]	89
	[1]	92
	[2]	97
	[3]	95
	[4]	90
	[5]	96
	[6]	94
	[7]	92
	[8]	90
	[9]	98
Largest		92
i		2
score[i]		97

Figure 1.24: Memory inspection of a one-dimensional array 5.

In Figure 1.25, we execute the Go command and the program pauses again. Now, i = 9 and score[9] has value 98, which is larger than the value of Largest, 97.

```
#include <stdio.h>
#define SIZE 10

int main(void)
{
    int score[SIZE]
       ={89,92,97,95,90,96,94,92,90,98};
    int i;
    int Largest =score[0];
    for ( i = 0; i < SIZE; i++ )
    {
        if (Largest < score[i])
            Largest=score[i];
    }

    return 0;
}
```

Watch		✕
Name		**Value**
⊟ score		0x0018ff20
[0]		89
[1]		92
[2]		97
[3]		95
[4]		90
[5]		96
[6]		94
[7]		92
[8]		90
[9]		98
Largest		97
i		9
score[i]		98

Figure 1.25: Memory inspection of a one-dimensional array 6.

In Figure 1.26, the loop terminates and the final value of Largest is 98.

```
#include <stdio.h>
#define SIZE 10

int main(void)
{
    int score[SIZE]
        = {89,92,97,95,90,96,94,92,90,98};
    int i;
    int Largest =score[0];
    for ( i = 0; i < SIZE; i++ )
    {
        if (Largest < score[i])
            Largest=score[i];
    }

    return 0;
}
```

Watch		✕
Name		**Value**
⊟ score		0x0018ff20
[0]		89
[1]		92
[2]		97
[3]		95
[4]		90
[5]		96
[6]		94
[7]		92
[8]		90
[9]		98
Largest		98
i		10
score[i]		1638280

Figure 1.26: Memory inspection of a one-dimensional array 7.

Example 1.2 Computing total score
Scores given by judges are stored in an array score[10].

[Analysis]
1.   Algorithm design
The algorithm is shown in Figure 1.27. Code implementation can be easily adapted from the pseudo code in the second refinement.

Top-level pseudo code	First refinement	Second refinement
Compute sum of scores stored in array score[10]	Use total to store the sum ,and score[10] to store scores	Initialize score[10]
		Sum total =0;
	Add values of elements in score to total repeatedly	while (i<10)
		total += score[i];
		i++;
	Output result	Output total

Figure 1.27: Computing total score.

After eliminating the highest score and the lowest score, we can compute the total score that complies with the scoring rule.

2.   Code implementation

```
01 //Compute sum of array elements
02 #include <stdio.h>
03 #define SIZE 10
04
05 int main(void)
06 {
07      int score[ SIZE ] = {98,92,89,95,90,96,94,92,90,97};
08      int i;                       //counter
09      int total = 0;               //sum
10
11      for ( i = 0; i < SIZE; i++ )
12      {
13        total +=score[ i ];      //Compute sum of array elements
14      }
15      printf( "The total score is %d\n", total );
16      return 0;
17 }
```

Program result:
The total score is 933

Example 1.3 Number guessing game

An array stores an increasing number sequence 5, 10, 19, 21, 31, 37, 42, 48, 50, 55. Use binary search to find elements with key values 19 and 66.

[Analysis]

1.   Algorithm analysis

Let low denote the position of the minimum value in the searching range, and high denote the position of the maximum value in the searching range. The comparison position in binary search is then mid = (low + high)/2. Comparing key value with the element at position mid yields one of the following results:

−   Equal: the element at position mid is what we are looking for.
−   Greater: we will look for the element in the lower range by setting low = mid + 1.
−   Less: we will look for the element in the higher range by setting high = mid − 1.

Figures 1.28 and 1.29 illustrate processes of finding values 19 and 66.

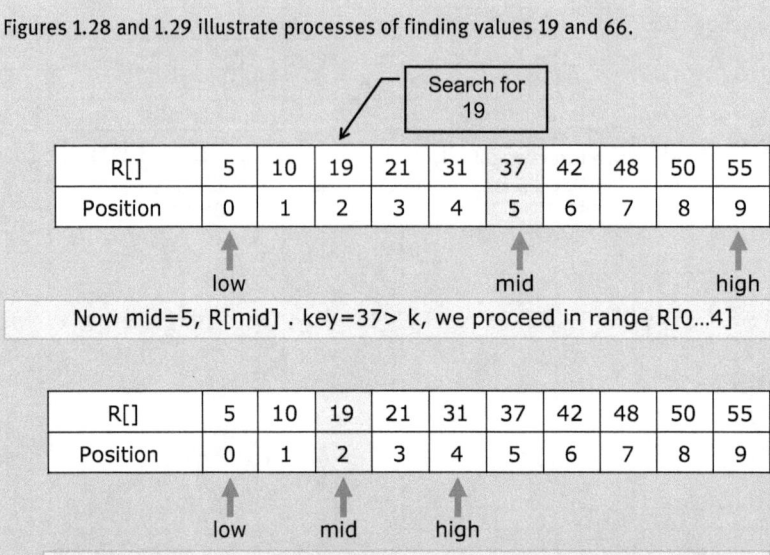

Figure 1.28: Binary search: searching for k = 19.

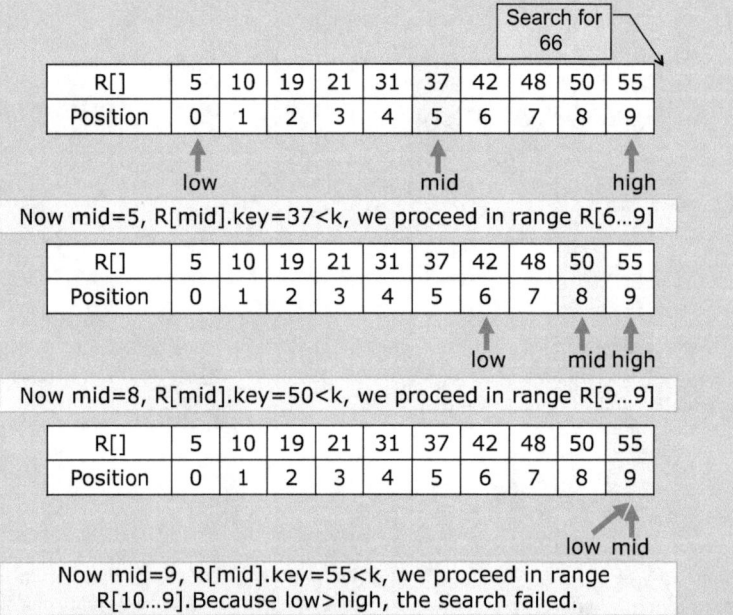

Figure 1.29: Binary search: searching for k = 66.

2. Code implementation

```c
#include <stdio.h>
#define N 10
int main(void)
{
  int a[N]={5,10,19,21,31,37,42,48,50,55};
  int low=0, high=N-1,mid;
  int key;
  int flag=0;                  //Search flag, 0=fail, 1=success
  printf("Please enter number to search:");
  scanf("%d",&key);
  while (low<=high)            //Search range is not empty
  {
  mid = (low+high+1)/2;
  if (a[mid]== key)            //Match
  {
    flag=1;
    break;
  }
  else
  {
    if (a[mid]> key) high = mid-1; //Continue searching in lower range
    else low = mid+1;              // Continue searching in higher range
  }
  }
  if (flag==1)
  printf("Search succeeded, index of %d is %d\n",key,mid);
  else
  printf("Search failed\n");
  return 0;
}
```

Example 1.4 Assign values to a one-dimensional array using loop
Find the first 20 entries of the Fibonacci sequence.
    The Fibonacci sequence is as follows: 0, 1, 1, 2, 3, 5, 8, 13, 21, 34, . . . Its recurrence equation is
$$F(0) = 0, \quad F(1) = 1,$$
$$F(n) = F(n-1) + F(n-2)$$

[Analysis]
1. Data structure design
Since indices and values in the Fibonacci sequence are 1-to-1 corresponded, we can store values into a one-dimensional integer array, which will be represented by int f[20] in this example.
2. Algorithm design
We shall construct the first 20 entries based on the recurrence equation of the Fibonacci sequence, store them into the array and eventually output them.

The algorithm is shown in Figure 1.30.

Top-level pseudo code	First refinement	Second refinement
Store result in an array of size 20 and output result	Initialize array [20] with the first two values of the sequence	int f[20]={ 0, 1 } i=2;
	Fill in the array using the recurrence relation starting from f[2]	while i< 20
		f [i] = f [i-1]+f [i-2];
		i++;
	Output result	Output elements in array f

Figure 1.30: Computing values of Fibonacci sequence.

3.   Code implementation

```
1   //Find first 20 entries in Fibonacci sequence
2   #include <stdio.h>
3   int main(void)
4   {
5     int i;
6     int f[20]={0, 1}; //Array initialization
7
8     for (i=2; i<20; i++) //Generate the sequence
9     {
10      f[i]=f[i-1]+f[i-2]; //Recurrence equation of Fibonacci equation
11    }
12    for (i=0; i<20; i++) //Output array elements
13    {
14      if (i%5==0) printf("\n"); //Print 5 entries on each line
15      printf("%8d", f[i]);
16    }
17    return 0;
18 }
```

Program result:
```
   0     1     1     2     3
   5     8    13    21    34
  55    89   144   233   377
 610   987  1597  2584  4181
```

4.   Program analysis

We shall analyze characteristics of iterated data processing by reading the program.

Lines 8–11 insert values into the Fibonacci array. Let the index be i, which corresponds to array element f[i]. We can construct a table for them and fill in it with their values, as shown in Figure 1.31. In addition to dynamic tracing and debugging, a static approach like this can also help us analyze patterns in program execution. Note that indices start from 0, so the index of the last element should be one less than the array size.

Index i	0	1	2	3	4	5	...	18	19	20
f[i]	0	1	2	3	5	8	...	...	...	...

Figure 1.31: Analysis of Fibonacci sequence program.

5. Discussion
    (1) What if we do not initialize array f?
         Discussion: If so, values of f[0] and f[1] will be arbitrary values, so further computation will be wrong.
    (2) How can we construct the Fibonacci sequence of arbitrary size?
         Discussion: We can make the array size a symbol constant, so the program can be easily adapted.
    (3) What if we change the execution condition of the first for loop (line 8) to i ≤ 20?
         Discussion: An out-of-bound error will happen because we are going to write to f[20], which is not in the range of the array. This is a logic error in the program.

Program reading exercise
Teacher review system statistics
The university Mr. Brown works for has built an online teacher review system, where students can rate teachers with a score in the range 6–10. Suppose we have randomly collected 50 ratings of a teacher and stored them into an array, please write a program that generates number of occurrences of each possible score.

1. Algorithm description
Let the ratings array be rating[]. It records number of occurrences of each score. The index i can be computed by subtracting 6 from score x ($6 \leq x \leq 10$), that is, $i = x-6$, so we can use values score−6 as indices of the ratings array. Whenever we find a new occurrence of a certain score, we add one to the corresponding array element.

2. Code implementation

```
1   #include<stdio.h>
2   #define RESPONSE_NUM 50      //Size of review array
3   #define RATING_SIZE 5          //Size of ratings array
4
5   int main(void)
6   {
7     int answer;              //Counter
8     int counter;
9
10    int rating[RATING_SIZE]={0};     //Rating array
11    int responses[RESPONSE_NUM]      //Review array that stores students' reviews
12    ={ 6,8,9,10,6,9,8,7,7,10,6,9,7,7,7,6,8,10,7,
13     10,8,7,7,6,7,8,9,7,8,7,10,6,7,6,7,7,10,8,
14     6,7,7,8,6,6,7,8,9,7,7,10
15    };
16
```

```
17   //Use score-6 as index of rating array, add 1 to an element if we find new
18   //occurrence of the corresponding score
19   for (answer=0; answer<RESPONSE_NUM; answer++)
20   {
21    rating[ responses[answer]-6 ]++;
22   }
23
24   //Print result in a table
25   printf("%s%17s\n","Rating","Number of occurrences");
26   for (counter=0; counter<RATING_SIZE; counter++)
27   {
28   printf("%6d%17d\n",counter+6,rating[counter]);
29   }
30   return 0;
31 }
```

Program result:

Rating	Number of occurrences
6	10
7	19
8	9
9	5
10	7

1.4 Operations on two-dimensional arrays

Having seen operations on one-dimensional arrays, we can proceed to two-dimensional arrays.

Example 1.5 Finding maximum in a two-dimensional array
There were three groups in Mr. Brown's class, each with six students. Now that the final exam has finished, please write a program to find the highest score and the corresponding student.

[Analysis]
1.   Data description
As shown in Figure 1.32, we can store the scores in a two-dimensional array.

Essentially, this problem is equivalent to finding the maximum value in a two-dimensional array with N rows and M columns and its row and column indices. To do this, we can simply repeat the process of finding the maximum value in a one-dimensional array N times.

Figure 1.33 shows how row and column indices change when traversing the array in a row-first manner. We first traverse row 0, with column index changing from 0 to M−1. Then we traverse row 1, with column index changing from 0 to M−1 as well. We repeat this process until we reach row N−1.

Group	Grade					
1	80	77	75	68	82	78
2	78	83	82	72	80	66
3	73	50	62	60	91	72

Figure 1.32: Exam results.

Column / Row	0	1	2	3	4	5
0	80	77	75	68	82	78
1	78	83	82	72	80	66
2	73	50	62	60	91	72

	Changes of row and column values when traversing in a row-first manner			
Row i	0	1	...	N-1
Column j	0~M-1	0~M-1	...	0~M-1

Figure 1.33: Traversing order of two-dimensional arrays.

2. Algorithm description
Figures 1.34 and 1.35 show the pseudo code of the algorithm.

Top-level pseudo code	First refinement
Input 2-d array	Input 2-d array
Find the maximum element and its row and column indices	Use the first element as comparison basis max
	Compare each element (row-first manner)with max, Update max with the larger Record the corresponding indices line and col
Output result	Output result

Figure 1.34: Pseudo code of finding maximum value in two-dimensional array 1.

Second refinement	third refinement

Input 2-d array a[N][M] in a row-first manner(or initialize)	int i, j, a[N][M], max, line, col;
	for(i=0;i<N;i++)
	   for(j=0;j<M;j++)
	      scanf("%d", &a[i][j]);

max=a[0][0];　　line=col=0;	max=a[0][0];　　line=col=0;
i=j=0;	
while row index i<N	for(i=0;i<N;i++)
while column index j<M	for(j=0;j<M;j++)
if (max<a[i][j])	if (max<a[i][j])
max=a[i][j]	{
line=i	max=a[i][j];
col=j	line=i;
j++;	col=j;
i++;　j=0;	}
Output max,line and col	printf("\n max=%d\t line=%d\t col=%d\n", max, line, col);

Figure 1.35: Pseudo code of finding maximum value in two-dimensional array 2.

3. Code implementation

We can write the code based on the second refinement, in which we use for statements to implement while loops. The complete code is as follows:

```
01 #include <stdio.h>
02 #define N 3
03 #define M 6
04
05 int main(void)
06 {
07   int i,j,max,line,col;
08   int a[N][M]= { {80,77,75,68,82,78},
09                  {78,83,82,72,80,66},
10                  {73,50,62,60,91,72}
11                };
12   max=a[0][0];
13   line=col=0;
14   for (i=0; i<N; i++)
15   {
16       for ( j=0; j<M; j++)
17       {
18            if (max<a[i][j])
19            {
20                max=a[i][j];
21                line=i;
22                col=j;
23            }
24       }
25   }
26   printf("max=%d\t line=%d\t col=%d\n",max,line,col);
```

```
27  return 0;
28 }
```

Program result:

```
max=91  line=2  col=4。
```

4. Debugging

Based on the characteristics of two-dimensional arrays and key points of this problem, we designed a few test cases for debugging, as shown in Figure 1.36.

Debugging plan
• Inspect memory layout of 2-d array
• Pattern of row and column indices
• Use breakpoints to find required values quickly

```
18          if(max<a[i][j])
19          {
20 ●            max=a[i][j];
21              line=i;
22              col=j;
23          }
```

Figure 1.36: Key points of debugging the program that finds maximum value in a two-dimensional array.

One may notice that the row addresses of a two-dimensional array are represented in the form of a one-dimensional array in the IDE debugger, as shown in Figure 1.37. To traverse the entire array, we traverse every column for each row. Note that a two-dimensional array is stored row by row in memory (each row as a one-dimensional array).

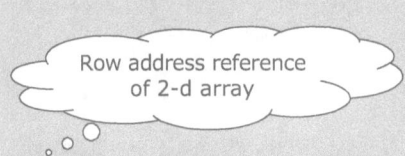

Row address reference of 2-d array

	Address	Row	0	1	2	3	4	5
a[0]	0x18feec	0	80	77	75	68	82	78
a[1]	0x18ff04	1	78	83	82	72	80	66
a[2]	0x18ff1c	2	73	50	62	60	91	72

Pattern of row and column indices				
Row I	0	1	...	N-1
Column j	0~M-1	0~M-1	...	0~M-1

Watch	
Name	**Value**
⊟ a	0x0018feec
⊟ [0]	0x0018feec
[0]	80
[1]	77
[2]	75
[3]	68
[4]	82
[5]	78
⊟ [1]	0x0018ff04
[0]	78
[1]	83
[2]	82
[3]	72
[4]	80
[5]	66
⊟ [2]	0x0018ff1c
[0]	73
[1]	50
[2]	62
[3]	60
[4]	91
[5]	72

Figure 1.37: Data storage in finding maximum in two-dimensional array problem.

As shown in Figure 1.38, we insert one breakpoint to the line where the current maximum value is updated and to the line where the result gets printed. When the program enters the first loop, as shown in Figure 1.39, the 0th element of the array is selected as the comparison basis, whose value is a[0][0] = 80. In Figure 1.40, the program pauses after we execute the Go command. The value of the element with index i = 0 and j = 4 is 82, which is larger than max.

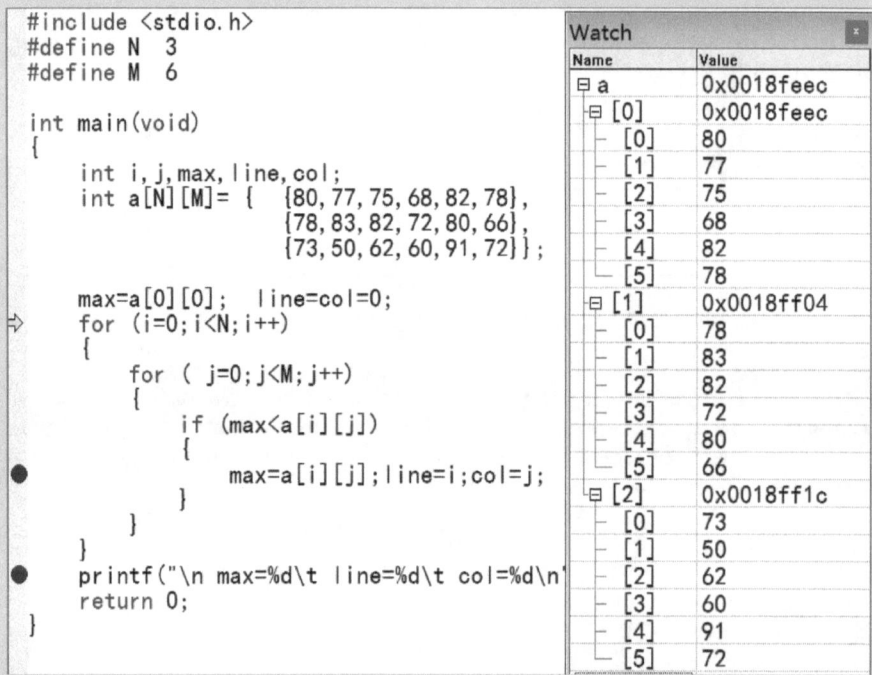

Figure 1.38: Debugging the program that finds maximum in two-dimensional array 1.

```
    for (i=0; i<N; i++)
    {
        for ( j=0; j<M; j++)
        {
            if (max<a[i][j])
            {
                max=a[i][j]; line=i; col=j;
            }
        }
```

Watch		
Name	Value	
max	80	
a[i][j]	80	
i	0	
j	0	

Figure 1.39: Debugging the program that finds maximum in two-dimensional array 2.

```
for (i=0; i<N; i++)
{
    for ( j=0; j<M; j++)
    {
        if (max<a[i][j])
        {
            max=a[i][j];line=i;col=j;
        }
    }
}
```

Watch	
Name	Value
max	80
a[i][j]	82
i	0
j	4

Figure 1.40: Debugging the program that finds maximum in two-dimensional array 3.

In Figure 1.41, the program pauses after we execute the Go command. The value of the element with index i = 1 and j = 1 is 83, which is larger than max.

```
for (i=0; i<N; i++)
{
    for ( j=0; j<M; j++)
    {
        if (max<a[i][j])
        {
            max=a[i][j];line=i;col=j;
        }
    }
}
```

Watch	
Name	Value
max	82
a[i][j]	83
i	1
j	1

Figure 1.41: Debugging the program that finds maximum in two-dimensional array 4.

In Figure 1.42, the program pauses after we execute the Go command. The value of the element with index i = 2 and j = 4 is 91, which is larger than max.

```
for (i=0; i<N; i++)
{
    for ( j=0; j<M; j++)
    {
        if (max<a[i][j])
        {
            max=a[i][j];line=i;col=j;
        }
    }
}
```

Watch	
Name	Value
max	83
a[i][j]	91
i	2
j	4

Figure 1.42: Debugging the program that finds maximum in two-dimensional array 5.

In Figure 1.43, the program completed scanning the array, and the loop is terminated. Now, i = 3, j = 6, and the maximum value of the array is max = 91.

```
for (i=0; i<N; i++)
{
    for ( j=0; j<M; j++)
    {
        if (max<a[i][j])
        {
            max=a[i][j];line=i;col=j;
        }
    }
}
printf("\n max=%d\t line=%d\t col=%d\n"
```

Watch		
Name	Value	
max	91	
a[i][j]	4199033	
i	3	
j	6	

Figure 1.43: Debugging the program that finds maximum in two-dimensional array 6.

Conclusion Execution order of nested loops
As shown in Figure 1.44, C has the following rules for executing nested loops:
1. Check the outer loop execution condition: if it is met, the body of the outer loop is executed; otherwise, the outer loop is terminated.
2. Check the inner loop execution condition: if it is met, the body of the inner loop is executed; otherwise, the inner loop is terminated, and the program proceeds to loop increment of the outer loop.

Program reading exercise Whac-A-Mole
Whac-A-Mole is a classic computer game, in which moles pop up from holes at random. Players need to force them back to their holes and obtain rewards by using a mallet to hit the moles on the head.

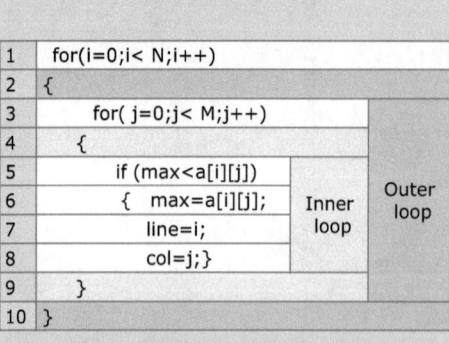
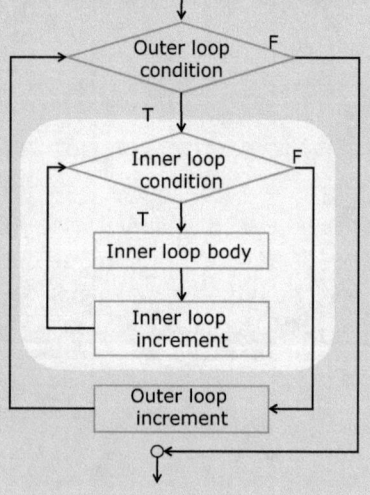

Figure 1.44: Execution order of nested loops.

1. Algorithm description

The program uses random functions srand and rand to generate positions at which moles appear. The following Whac-A-Mole program has a 3 by 3 "ground" and treats user input coordinates as positions the mallet hits. Although it is a console program and has a simple user interface, the way it works is the same as a Whac-A-Mole game with beautiful graphics.

2. Code implementation

```c
#include <stdio.h>
#include <stdlib.h>
#include <time.h>
//To simplify the code, we omitted curly brackets for some of the if-else statements
int main(void)
{
  int times = 0;     //Number of chances
  int mousey = 0;   //Row index of the mole
  int mousex = 0;   //Column index of the mole
  int posy = 0;      //Row index of the mallet
  int posx = 0;     //Column index of the mallet
  int hits = 0;     //Number of hits
  int missed = 0;   //Number of misses
  int num = 0, row = 0, col = 0;
  srand(time(0));
  //Obtain game chances
  printf("How many times do you want to play? : ");
  scanf("%d", &times);
  //Print the map
  printf("***\n***\n***\n");
  printf("Mallet position input should be row index followed by column index, separated by
space\n");
  //Actual game process
  for (num = 1; num <= times; num++)
  {
  //Obtain position of mole and mallet
          mousey = rand() % 3 + 1;
  mousex = rand() % 3 + 1;
  do
   {
  printf("Enter mallet position: ");
  scanf("%d %d", &posy, &posx);
  } while (posy < 1 || posy > 3 || posx < 1 || posx > 3);
  //Update number of hits and misses
  if (mousey == posy && mousex == posx) hits++;
  else missed++;
  //Print the map
          for (row = 1; row <= 3; row++)
  {
  for (col = 1; col <= 3; col++)
  {
```

```
if (row == posy && col == posx) printf("O");
else if (row == mousey && col == mousex) printf("X");
else printf("*");
}
printf("\n");
}
//Text indicating hit or miss
if (mousey == posy && mousex == posx) printf("Bingo!\n");
else printf("You missed.\n");
    //Print total score
printf("%d hits, %d misses\n", hits, missed);
}
return 0;
}
```

Program reading exercise Determining nationality
Six people are staying at a hotel, each from a different country. These countries are America, Germany, Britain, France, Russia, and Italy. We shall use letters A to F to denote these people. It is known that:
(1)   A and the American are doctors.
(2)   E and the Russian are technicians.
(3)   C and the German are technicians.
(4)   B and F used to be soldiers, and the German has never been a soldier.
(5)   The French is older than A; the Italian is older than C.
(6)   B and the American are going to Xi'an next week, while C and the French are going to Hangzhou next week.

Determine the nationalities of these people.

[Analysis]
1.   Data analysis
We shall first use the given information to eliminate the wrong answers.
     Based on conditions 1, 2, and 3, we can conclude that A is not American, E is not Russian, and C is not German. Based on occupation limits (A and the German have different jobs, so do E and the American, E and the German, C and the American, and C and the Russian), it is clear that A is neither Russian nor German, E is neither American nor German, and C is neither American nor Russian.
     It can be inferred from conditions 4 and 5 that neither B nor F is German, A is not French, and C is not Italian.
     Given condition 6, we know B is neither American nor French (because B and the French are going to different cities next week), and C is not French.
     To sum up:
     A: A is not American, Russian, German, or French.
     B: B is not German, American, or French.
     C: C is not German, American, Russian, Italian, or French.
     D: no information.
     E: E is not American or German.
     F: F is not German.

We can store the earlier information into matrix a, and country names into another one-dimensional array countries, as shown in Figure 1.45.

Rows of matrix a represent these guests, while columns represent their home countries. The 0th row is a special row for progress flags, which is either 1 for not processed or 0 for processed. The values of other elements indicate nationalities. For example, 4 represents Germany in the countries array. If a value is 0, the person represented by the row does not come from the country represented by the column.

2. Algorithm design

Following steps 2 and 3 in Figure 1.45, we can find the solution by repeatedly zeroing out rows.

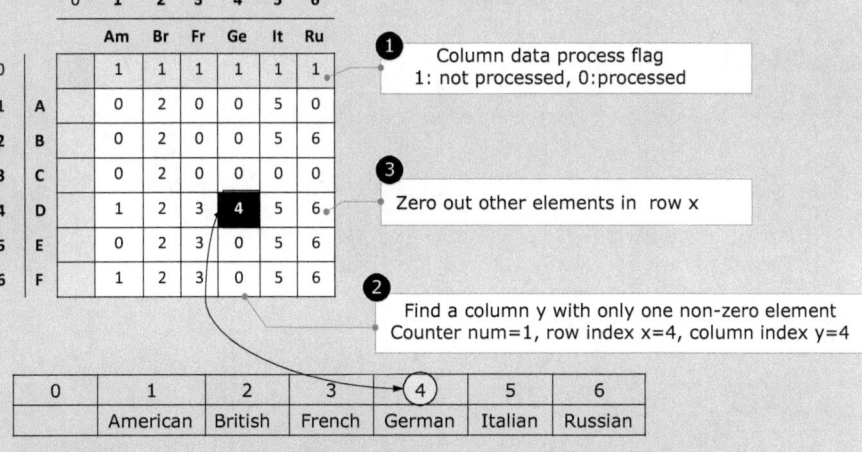

*countries[7] = {" ","American","British","French","German","Italian","Russian "}; }

Figure 1.45: Storage and procedures of determining nationalities problem.

3. Code implementation

```
#include<stdio.h>
char *countries[7]={" ","American","British","French","German","Italian","Russian"};
//The asterisk before countries indicates that the array stores addresses,
//which are beginning addresses of strings
int main(void)
{
  int a[7][7],i,j,k,num,x,y;
  for(i=0;i<7;i++) //Initialize the matrix
  for(j=0;j<7;j++) a[i][j]=j; //Row for person, column for country,
                  //and value for nationality
  for(i=1;i<7;i++) a[0][i]=1; //0-th element in each column is the progress mark,
                  //1 means not processed
  //Enter know information, 0 means the person is not from a country
  a[1][1] = a[1][3] = a[1][4] = a[1][6] = 0; // A is not American, Russian, German or French
  a[2][1]= a[2][3]= a[2][4] =0; // B is not German, American or French
  a[3][1] = a[3][3] = a[3][4]= a[3][5] =a[3][6] = 0;
  // C is not German, American, Russian, Italian or French
```

```
a[5][1] = a[5][4]= 0; // E is not American or German
a[6][4]=0; //F is not German
while(a[0][1]+a[0][2]+a[0][3]+a[0][4]+a[0][5]+a[0][6]>0)
//Jump out of the loop once every column is processed
{
 for(i=1;i<7;i++) //i is column index, we process the matrix column by column
 {
  if(a[0][i]) //Process the column if it hasn't been processed
  {
   for(num=0,j=1;j<7;j++) //j is row index
   {
    if(a[j][i])
    {
     num++; //num counts non-zero elements in the column
     x=j;
     y=i; //x and y are coordinates of the non-zero element
    }
   }
   if(num==1) //If there is only one non-zero element,
   //zero out the row (except the non-zero element)
   {
    for(k=1;k<7;k++)
    {
     if(k!=y)a[x][k]=0;
     a[0][y]=0; //Set column y to be "processed"
    }
   }
  }
 }
}
for(i=1;i<7;i++) //Print result
{
 printf("%c is", 'A'-1+i); //Print person
 for(j=1;j<7;j++)
 {
  if(a[i][j]!=0)
  {
   printf("%s\n",countries[a[i][j]]); //Print country
   break;
  }
 }
}
return 0;
}
```

Program result:

A is Italian
B is Russian

```
C is British
D is German
E is French
F is American
```

1.5 Operations on character arrays

Example 1.6 Password verification

When a user logs into a system, the system needs to compare the password he/she enters with the one used for registration. For example, a user signed up with password abc24680, as shown in Figure 1.46. How should the system store this password?

Index	0	1	2	3	4	5	6	7	8	9	...	18	19
Registered password	'a'	'b'	'c'	'2'	'4'	'6'	'8'	'0'					

Figure 1.46: Password used for registration.

[Analysis]

1. Storage structure of data

If we use character arrays to store passwords, there are two possible ways to assign initial values: the first is to assign characters one by one, while the other is to assign a string. Characters stored in these two approaches are the same, but termination mark '\0' will be automatically inserted to the end of the string by the system, as shown in Figure 1.47.

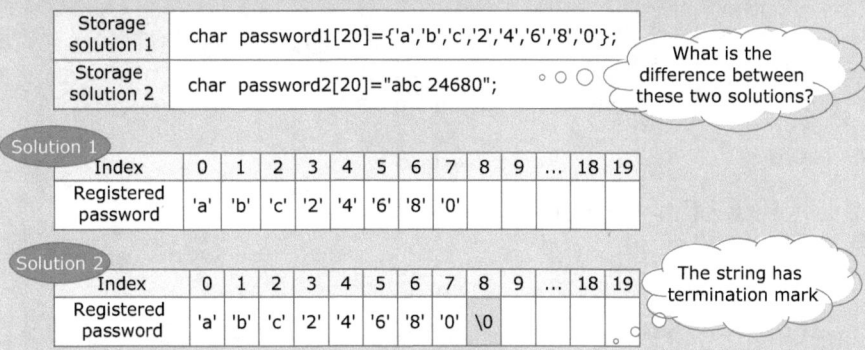

| Storage solution 1 | char password1[20]={'a','b','c','2','4','6','8','0'}; |
| Storage solution 2 | char password2[20]="abc 24680"; |

What is the difference between these two solutions?

Solution 1

Index	0	1	2	3	4	5	6	7	8	9	...	18	19
Registered password	'a'	'b'	'c'	'2'	'4'	'6'	'8'	'0'					

Solution 2

Index	0	1	2	3	4	5	6	7	8	9	...	18	19
Registered password	'a'	'b'	'c'	'2'	'4'	'6'	'8'	'0'	\0				

The string has termination mark

Figure 1.47: Storage approaches of character sequence.

Note: one can store strings of any length in C. When storing strings in character arrays, programmers need to make sure that the array size is large enough so that the longest string can fit in; if the string is longer than the array, characters beyond the array bound will override data after the array in memory.

2. Algorithm description

Figure 1.48 shows the stepwise refined algorithm.

Top-level pseudo code	First refinement
Compare keyboard input characters and registered password characters one by one Output "Password is wrong" upon mismatch	Store registered password in array password[ ]
	Read keyboard input character in ch

	while there is remaining input
	    Compare ch with password[ ]
	    Output "Password is wrong" upon mismatch

Output "Password is correct" if there is no mismatch	Output "Password is correct" if there is no mismatch

Second refinement
char password[20]; int i=0;
ch=getchar();

while (ch!='\n')
        if (ch != password[i])
            printf("Password is wrong");
            Jump out of loop
        ch=getchar();
        i++;

if (i==strlen(password))   printf("Password is correct");

Figure 1.48: Password verification.

In the second refinement, ch! = '\n' checks whether there are more inputs. The loop control variable i acts as a counter as well. strlen is a library function that computes string length (not counting termination mark '\0'). To determine whether the entire string has been checked, we compare i with the string length.

3.   Code implementation

```
01 #include <stdio.h>
02 #include <string.h>
03 int main(void )
04 {
05   int i=0;
06   char ch;
07   char password[20]="abc24680";
08   ch=getchar();
09   while (ch!='\n')
10   {
11     if (ch != password[i]) break;
12     ch=getchar();
13     i++;
14 }
15   if (i==strlen(password)) printf("Password is correct\n");
16   else printf("Password is wrong\n");
17   return 0;
18 }
```

Note that the header file for library function strlen on line 19 is included on line 2.

Example 1.7 Cracking Caesar code
What did the mysterious email Daniel sent to his father (see Figure 1.49) say? How many characters were shifted? How should Mr. Brown implement his algorithm?

ciphertext[]="lettc fmvxlhec hehhc pszi csy"

Ciphertext ——→ Decryption ——→ Plaintext

Left shift by ? positions

Figure 1.49: Cracking Caesar code.

[Analysis]
1.   Data processing
Without loss of generality, we shall use right shift (the alphabet is shifted by one character to its right each time) in the following discussion. To crack the ciphertext, we can list all 26 possible results and look for a meaningful string. Figure 1.50 shows the case of shifting by one character.

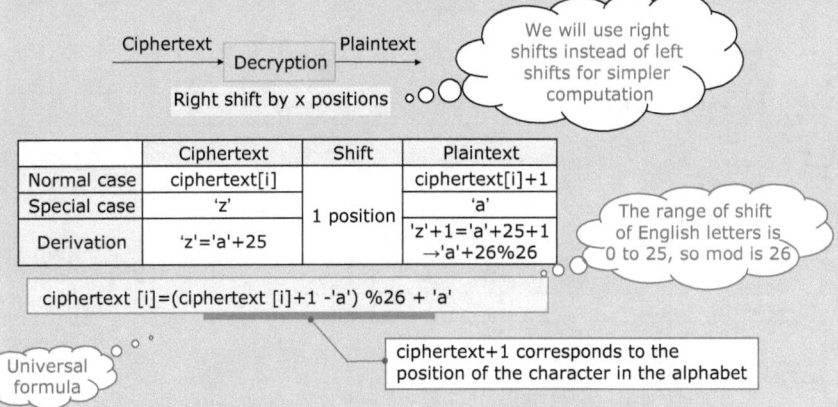

Ciphertext ——→ Decryption ——→ Plaintext

Right shift by x positions

We will use right shifts instead of left shifts for simpler computation

	Ciphertext	Shift	Plaintext
Normal case	ciphertext[i]		ciphertext[i]+1
Special case	'z'	1 position	'a'
Derivation	'z'='a'+25		'z'+1='a'+25+1 →'a'+26%26

The range of shift of English letters is 0 to 25, so mod is 26

ciphertext [i]=(ciphertext [i]+1 -'a') %26 + 'a'

Universal formula

ciphertext+1 corresponds to the position of the character in the alphabet

Figure 1.50: Character shifting analysis.

Normally, if ciphertext is ciphertext[i], its plaintext would be ciphertext[i] + 1, except character 'z', whose plaintext is 'a'. In other words, we need to return to the beginning of the alphabet when reaching the end. Let us examine this case more carefully.

Ciphertext character 'z' can be represented by the character 'a' plus 25, namely 'z' = 'a' + 25. Hence plaintext 'z' + 1 = 'a' + 25 + 1 should be the character 'a'.

We can use modular arithmetic (mod 26) to eliminate the 26 in the equation. Modular arithmetic helps us return to the beginning of the alphabet when going out of bound.

By now, we have derived the universal formula for right shifting by 1 character. The expression inside parentheses indicates the position of ciphertext character plus 1 in the alphabet. For example, if character ciphertext[i] is 'b', we have:

ciphertext[i]+1-'a'='b'+1-'a'=2 //'b'+1 is shifted by 2 characters in the alphabet
(ciphertext[i]+1-'a')%26+'a'=2%26+'a'='c' //'b' becomes 'c' after right shifting by 1
character

2. Algorithm description

Figure 1.51 shows the pseudo code of the algorithm.

In the second refinement, '\0' is used to determine whether the entire string has been processed. Space is represented by a space wrapped with single quotation marks. The shifted ciphertext is computed using the formula we derived earlier. When printing strings, a number indicating the number of characters shifted is added to the beginning. Finally, we need to find a meaningful string in printed contents manually.

3. Code implementation

```
01 #include "stdio.h"
02 #define SIZE 80
03 int main(void)
04 {
05   char ciphertext[SIZE]="lettc fmvxlhec hehhc pszi csy";
06   int i=0,j=0;
07   printf( "%s\n",ciphertext);
08   while (j<26)
09   {
```

Top-level pseudo code	First refinement
	Repeat process below 26 times
– Right shift string by 1 position – Print string – Repeat process above 26 times	while not reaching string end ciphertext is not space right shift ciphertext print ciphertext string

Second refinement
while(j<26)
while(ciphertext[i]!='\0')   //while not reaching string end
if (ciphertext[i] !=' ')   //skipspace
ciphertext[i]=(ciphertext[i]+1-'a')%26+'a' //right shift by 1
i++
printf("%d:%s\n", j, ciphertext)
i=0
j++

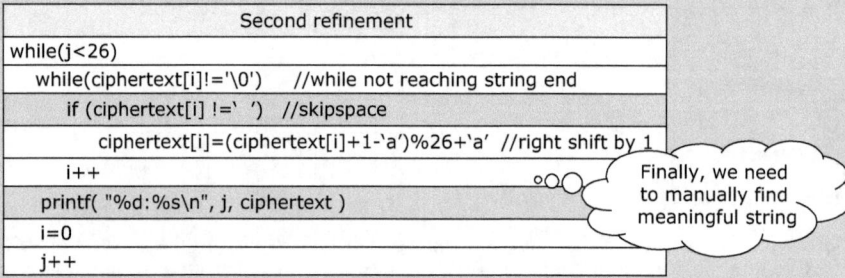

Finally, we need to manually find meaningful string

Figure 1.51: Algorithm for cracking Caesar code.

```
10   while (ciphertext[i]!='\0')
11   {
12    if (ciphertext[i]!=' ')
13    {
14     ciphertext[i]=(ciphertext[i]+1-'a')%26+'a';
15    }
16    i++;
17   }
18   printf("%d:%s\n",j,ciphertext);
19   i=0;
20   j++;
21  }
22  return 0;
23 }
```

Program result:
```
lettc fmvxlhec hehhc pszi csy
0:mfuud gnwymifd ifiid qtaj dtz
1:ngvve hoxznjge jgjje rubk eua
2:ohwwf ipyaokhf khkkf svcl fvb
3:pixxg jqzbplig lillg twdm gwc
4:qjyyh kracqmjh mjmmh uxen hxd
5:rkzzi lsbdrnki nknni vyfo iye
6:slaaj mtcesolj olooj wzgp jzf
7:tmbbk nudftpmk pmppk xahq kag
8:unccl oveguqnl qnqql ybir lbh
9:voddm pwfhvrom rorrm zcjs mci
10:wpeen qxgiwspn spssn adkt ndj
11:xqffo ryhjxtqo tqtto belu oek
12:yrggp szikyurp uruup cfmv pfl
13:zshhq tajlzvsq vsvvq dgnw qgm
14:atiir ubkmawtr wtwwr ehox rhn
15:bujjs vclnbxus xuxxs fipy sio
16:cvkkt wdmocyvt yvyyt gjqz tjp
17:dwllu xenpdzwu zwzzu hkra ukq
18:exmmv yfoqeaxv axaav ilsb vlr
19:fynnw zgprfbyw bybbw jmtc wms
20:gzoox ahqsgczx czccx knud xnt
21:happy birthday daddy love you
22:ibqqz cjsuiebz ebeez mpwf zpv
23:jcrra dktvjfca fcffa nqxg aqw
24:kdssb eluwkgdb gdggb oryh brx
25:lettc fmvxlhec hehhc pszi csy
```

It is clear that the twenty-first string is what we want: "happy birthday daddy love you". Mr. Brown was impressed by what he saw: Daniel had not learned to program, but he was able to manually compute the ciphertext without mistakes.

Example 1.8 Sorting family names
Please write a program to sort the following family names in alphabetical order.
Zhao, Zhou, Zhang, Zhan, Zheng

[Analysis]

1.   Data storage
Each family name is a string, so multiple family names are multiple strings, which can be stored into a two-dimensional character array, as shown in Figure 1.52. Since we have five family names, the number of rows in the array should be 5. The longest name has five characters, so the number of columns should be 6 to store the name and a termination mark. The one-dimensional form of the array can represent the beginning address of a row in a two-dimensional array.

char c[3][6]={"Zhao","Zhou", "Zhang","Zhan","Zheng"}

C[0]	Z	h	a	o	\0	\0
C[1]	Z	h	o	u	\0	\0
C[2]	Z	h	a	n	g	\0
C[3]	Z	h	a	n	\0	\0
C[4]	Z	h	e	n	g	\0

The beginning address of a row in a 2-dimensional array can be represented by the 1-dimensional form of the array.

Figure 1.52: Storage of multiple strings.

2.   Algorithm description
The pseudo code is shown in Figure 1.53.
    C provides many library functions to process strings. As shown in Figure 1.54, this algorithm requires strcpy for string copying and strcmp for string comparison.

Top-level pseudo code	First refinement	Second refinement
	Store M strings in c[M][6]	char c[M][6],char str[6]
	Use the first string as comparison basis str	Use c[0] as comparison basis, and copy it into str
Find the largest string among multiple strings	Compare each string with in row order, store the larger in str	i=1;
		while i< M
		if str<c [i]
		Copy c[i]into str
		i++;
Output result	Output result	Output str

Figure 1.53: Sorting multiple strings.

Function	Functionality	Return value
strcpy(character array, string)	Copy string into the character array	
strcmp(string1,string2)	Compare two strings alphabetically	0:equal
		Positive number: string 1>string2
		Negative number: string 1<string2

#include <string.h>

Figure 1.54: String processing functions.

Knowledge ABC String processing functions

C provides various string processing functions, which handle input, output, concatenation, modification, comparison, conversion, copy, and search of strings. Using these functions simplifies programming tasks.

To use string functions for input and output, we should include the header file "stdio.h" first. To use other string functions, we should include the header file "string.h".

Please refer to Appendix C of Volume 1 for function prototypes and explanations of common string functions.

3. Code implementation

```
01 #include <stdio.h>
02 #include <string.h>
03 #define M 5
04 int main(void)
05 {
06   char c[M][6]={"Zhao","Zhou","Zhang","Zhan","Zheng"};
07   char str[6];
08   int i;
09
10   strcpy(str, c[0]); //Use string copy function to copy c[0] into array str
11   for (i=1; i<M; i++)
12   {
13     if (strcmp(str, c[i])< 0) //if str is less than c[i]
14     {
15       strcpy(str, c[i]); //then copy c[i] into str
16     }
17   }
18   printf("The largest string is:%s\n", str);
19   return 0;
20 }
```

Program result: The largest string is Zhou.

On line 10, we use the string copy function to copy c[0] into array str.
On line 13, we compare the contents of str and c[i].
On line 15, we copy the larger string into str.

4.   Debugging
Based on characteristics of the earlier program, we can conclude the key steps in debugging as follows:
Inspecting a two-dimensional character array: string initialization and termination mark.
Inspecting row addresses of a two-dimensional array: the beginning address of a row in a two-dimensional array is represented by the array name suffixed with one-dimensional index;
Inspecting how strcpy and strcmp functions work.
Figure 1.55 shows the two-dimensional character array c after initialization. Each row has length 6 and stores a string. If a string has less than six characters, the system pads it with 0. The address of a row is the beginning address of the string in that row, represented by c[i]. i is an integer in the range 0–5.

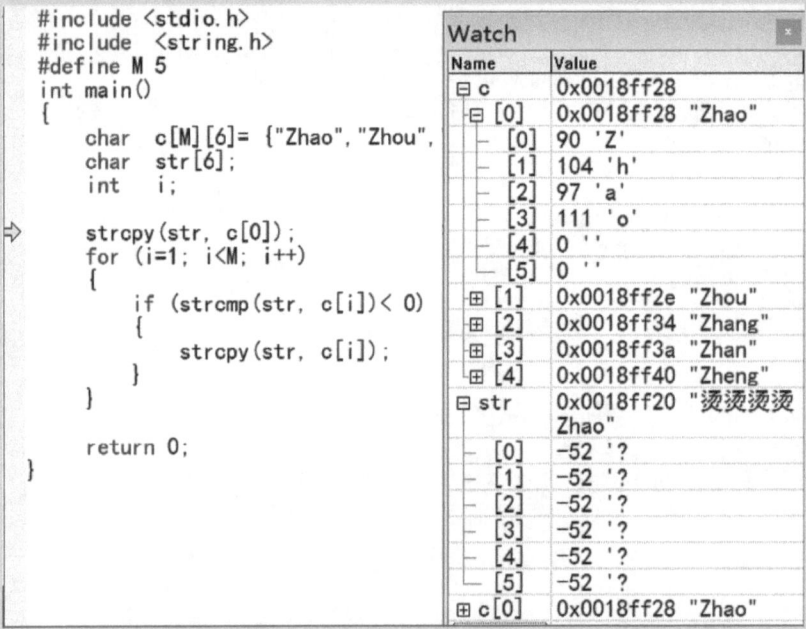

Figure 1.55: Finding the largest string debugging, step 1.

When the one-dimensional array character str was not initialized, its elements were decimal number −52, which corresponded to random Chinese characters. The reason "Zhou" is also displayed is that the system stops upon reaching the termination mark '\0' when displaying strings.

We use strcpy function to copy c[0] into str. In Figure 1.56, this change is shown in the Watch window.

```
#include <stdio.h>
#include <string.h>
#define M 5
int main()
{
    char  c[M][6]= {"Zhao", "Zhou",
    char  str[6];
    int   i;

    strcpy(str, c[0]);
    for (i=1; i<M; i++)
    {
        if (strcmp(str, c[i])< 0)
        {
            strcpy(str, c[i]);
        }
    }

    return 0;
}
```

Watch		☒
Name	Value	
⊞ c	0x0018ff28	
⊟ str	0x0018ff20 "Zhao"	
[0]	90 'Z'	
[1]	104 'h'	
[2]	97 'a'	
[3]	111 'o'	
[4]	0 ' '	
[5]	−52 '?	
▣ c[0]	0x0018ff28 "Zhao"	
[0]	90 'Z'	
[1]	104 'h'	
[2]	97 'a'	
[3]	111 'o'	
[4]	0 ' '	
[5]	0 ' '	

Figure 1.56: Finding the largest string debugging, step 2.

In Figure 1.57, i has value 1, and c[i] has value "Zhou" in the first iteration.

In Figure 1.58, the strcpy function in the if statement is executed and str now stores the string "Zhou". This indicates that the result of the strcmp function is less than 0.

1.6 Summary

Arrays are one of the most commonly used data structures in programming. An array can be one-dimensional, two-dimensional, or multidimensional.

An array declaration consists of a type identifier, an array name and an array length. An array element is also called an indexed variable.

Assigning values to an array can be done through initialization, input functions, or assignment statements. Figure 1.59 shows the use cases of these approaches.

The main contents and relations between them are shown in Figure 1.60.

A variable is a single datum,

Whereas an array stores a group of data together,

```
#include <stdio.h>
#include <string.h>
#define M 5
int main()
{
    char  c[M][6]= {"Zhao", "Zhou",
    char  str[6];
    int   i;

    strcpy(str, c[0]);
    for (i=1; i<M; i++)
    {
        if (stromp(str, c[i])< 0)
        {
            strcpy(str, c[i]);
        }
    }

    return 0;
}
```

Watch

Name	Value
⊞ c	0x0018ff28
⊟ str	0x0018ff20 "Zhao"
[0]	90 'Z'
[1]	104 'h'
[2]	97 'a'
[3]	111 'o'
[4]	0 ' '
[5]	-52 '?
⊟ c[0]	0x0018ff28 "Zhao"
[0]	90 'Z'
[1]	104 'h'
[2]	97 'a'
[3]	111 'o'
[4]	0 ' '
[5]	0 ' '

Figure 1.57: Finding the largest string debugging, step 3.

```
#include <stdio.h>
#include <string.h>
#define M 5
int main()
{
    char  c[M][6]= {"Zhao", "Zhou",
    char  str[6];
    int   i;

    strcpy(str, c[0]);
    for (i=1; i<M; i++)
    {
        if (stromp(str, c[i])< 0)
        {
            strcpy(str, c[i]);
        }
    }

    return 0;
}
```

Watch

Name	Value
⊟ str	0x0018ff20 "Zhou"
[0]	90 'Z'
[1]	104 'h'
[2]	111 'o'
[3]	117 'u'
[4]	0 ' '
[5]	-52 '?
i	1
⊟ c[i]	0x0018ff2e "Zhou"
[0]	90 'Z'
[1]	104 'h'
[2]	111 'o'
[3]	117 'u'
[4]	0 ' '
[5]	0 ' '

Figure 1.58: Finding the largest string debugging, step 4.

	Data characteristics	Use cases
Initialization	Can be either regular or not	We only need to type in data once. It is convenient to use initialization when data size is large and we need to debug repeatedly.
Keyboard input	Can be either regular or not	We need to type in data in every execution. Although input can vary, it is not convenient for debugging. We can use this method to test our program on different input after it is debugged.
Assignment statement	Regular	Values are assigned automatically. We can use this method when data are regular.

Figure 1.59: Assignment approaches and their use cases.

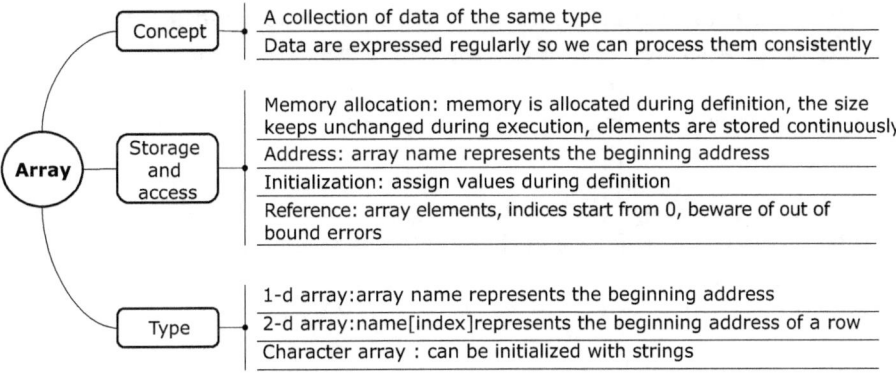

Figure 1.60: Relations between concepts related to arrays.

The three key elements of variables are name, value, and address,
Arrays are just about the same.
Memory is allocated to arrays during definition, and it does not change during execution,
An array name can also be used as the beginning address of the array,
Array elements are of the same type, but their values can be different.
An array element is similar to a variable,
The index indicates its position in the array,
We should remember that indices start at 0,
And that out-of-bound accesses lead to errors.
The system uses '\0' to mark the end of a character array.

1.7 Exercises

1.7.1 Multiple-choice questions

1. [Array definition]
   Which of the following statements define an array correctly? ( )
   A) int num[0. . .2008]
   B) int num[]
   C) int N=2008; int num[N]
   D) #define N 2008 int num[N];

2. [Character array]
   Which of the following statements is wrong about character arrays in C? ( )
   A) A character array can be used to store a string.
   B) A string stored in a character array can be input/output together.
   C) We can assign values to a character array using assignment operator "=" in an assignment statement.
   D) We cannot use relational operators to compare strings stored in character arrays.

3. [String assignment of character arrays]
   Suppose we have the following character array definition: char array[]="China";
   Then the size of the array is ( )
   A) 4 bytes   B) 5 bytes   C) 6 bytes   D) 7 bytes

4. [Character array: termination mark]
   Suppose we have the following character arrays: char x[]="abcdefg"; char y[]={'a','b','c','d','e','f','g'}; Which of the following statements is correct? ( )
   A) Array x is equivalent to array y.
   B) Array x and array y have the same length.
   C) Length of array x is larger than that of array y.
   D) Length of array x is smaller than that of array y.

5. [String input]
   Suppose we have char s[30]={0} and we type "This is a string. <Enter>" during program execution.
   Which of the following statements cannot read the entire string "This is a string". into character array s correctly? ( )
   A) i=0;while ((c=getchar())!='\n') s[i++]=c
   B) gets(s)
   C) for (i=0; (c=getchar()) !='\n'; i++) s[i]=c
   D) scanf("%s", s)

6. [Array access]
   What is the output of the following program? ( )

```
int y=18, i=0, j, a[8];
do
{
  a[i]=y%2;
  i++;
  y=y/2;
} while(y>=1);
for(j=i-1;j>=0;j--) printf("%d", a[j]);
```

   A) 10000   B) 10010   C) 00110   D) 10100

7. [Two-dimensional array]

```
int i, t[][3]={9,8,7,6,5,4,3,2,1};
for(i=0;i<3;i++) printf("%d ",t[2-i][i]);
```

   What is the output of the program above? ( )
   A) 3 5 7   B) 7 5 3   C) 3 6 9   D) 7 5 1

8. [Two-dimensional array]

```
int a[4][4]={ {1,4,3,2},{8,6,5,7},{3,7,2,5},{4,8,6,1}}, i, k, t;
for (i=0; i<3; i++)
for (k=i+1; k<4; k++)
if (a[i][i]<a[k][k])
{ t=a[i][i]; a[i][i]=a[k][k]; a[k][k]=t;}
for (i=0; i<4; i++)
printf("%d,", a[0][i]);
```

   What is the output of the program above? ( )
   A) 1,1,2,6,   B) 6,2,1,1,   C) 6,4,3,2,   D) 2,3,4,6,

9. [Characters in two-dimensional array]
   What is the output of the following program if we type in "peach flowers is pink
   <Enter>" during execution? ( )

```
char b[4][10]; int i;
for ( i=0; i<4; i++ ) scanf( "%s", b[i] );
for( i=3; i>=0; i-- ) printf( "%s ", b[i] );
```

A) peachflower is pink
B) pink is flower Peach
C) peachflowerispink
D) pink is flower peach

1.7.2 Fill in the tables

Complete tables in Figures 1.61–1.63 based on the program in each question.
1. [Two-dimensional array]

i	0	1	2	3
a[i][i]	1			None
s				End of loop
Output				

Figure 1.61: Arrays: fill in the tables, question 1.

	i	5	4	3	2	1	0
c[i]=c[i-1] Before	c[i]	Unknown					
assignment	c[i-1]	'\0'					
Output							

Figure 1.62: Arrays: fill in the tables, question 2.

	0	1	2	3	4	5	6	7	8	9
①p[ ]	a' '	b' '	c' '	d' '						
②p[ ]										
③p[ ]	'									
④Output										

Figure 1.63: Arrays: fill in the tables, question 3.

```
int main(void)
{
  int a[3][3]={1,2,3,4,5,6,7,8,9}, i, s=0;
  for (i=0; i<=2; i++)
    s=s+a[i][i];
  printf ( " s=%d\n " , s) ;
  return 0;
}
```

2. [One-dimensional character array]

```
int main(void)
{ int i=5;
  char c[6]="abcd";
  do
  {
    c[i]=c[i-1];
  } while(--i>0);
  puts(c);
  return 0;
}
```

3. [String processing library functions]

```
#include <string.h>
int main(void)
{
  char p[20]={'a', 'b', 'c', 'd'}; //——①
  char q[]="xyz", r[]="mnopq";
  strcat(p, r); //——②
  strcpy(p+strlen(q), q); //——③
  printf("%d\n", strlen(p)); //——④
  return 0;
}
```

1.7.3 Programming exercises

1. Thirteen people stand in a circle and are numbered off using only numbers 1, 2, and 3. That is, they shout out numbers 1, 2, 3, 1, 2, 3, 1, 2, 3. . . .. If someone shouts out number 3, that person should leave the circle. Write a program to find out which person is the last one remaining in the circle.

2. Write a program that reads a string from keyboard input, sorts it in ascending order based on ASCII values of characters, and prints the sorted string.

3. Please write a program, in which you define a one-dimensional character array str[50], read a sequence of characters from keyboard input and store it into str, read an integer M (M < 50), and finally copy characters after position M in array str into a new character array ch[50].

4. Write a program that finds the element with value x in a one-dimensional integer array with 10 elements. If such an element exists, the program should print its index; otherwise, the program should print "Search failed".

5. In an image encoding algorithm, we need to do a Zigzag scan on a given square matrix. The Zigzag scan processes of 4 × 4 and 5 × 5 matrices are illustrated in Figure 1.64. Write a program to simulate this process.

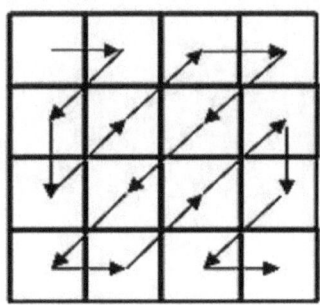 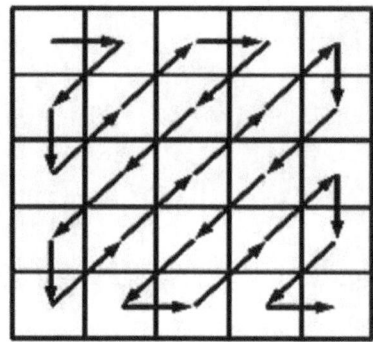

Figure 1.64: Zigzag scan.

2 Pointers

Main contents
- Meaning, usage, and examples of pointers
- Representation and nature of pointers, shown through comparing pointers with plain variables
- Differences and similarities between pointers and plain variables
- Relations between pointers and arrays
- Nature of pointer offsets
- Program reading practices
- Top-down algorithm design practices
- Debugging techniques of pointers

Learning objectives　　　　　　　　　　　　　　　　　　　　　　　　　　　　　**!**
- Understand the concept of pointers
- Understand relations between pointers, arrays, and strings
- Know how to use pointers to reference variables and arrays
- Can use string arrays through pointers
- Can design algorithms using the top-down stepwise refinement approach

2.1 Concept of pointers

2.1.1 Reference by name and reference by address

We shall explain these concepts through real-life examples.

Case study 1 Setting destination in a navigation system
Postmen nowadays often use navigation systems to find their destination when delivering to a new location. To set the destination in a navigation system, one can input either the address or name of the location. For example, "Xidian University North Campus" refers to the location by name, while "2 South Taibai Road, Yanta District" refers to the same location by address, as shown in Figure 2.1.

To sum up, we can reference an object that has a location attributing either by name or by address.

Case study 2 Classroom questioning and homework assignment
Teachers often ask students questions in class. He/she may ask "the second student in the third row" to answer the question when he/she does not know the name of the student. In this case, the student's name is a "reference by name," while the student's

https://doi.org/10.1515/9783110692303-002

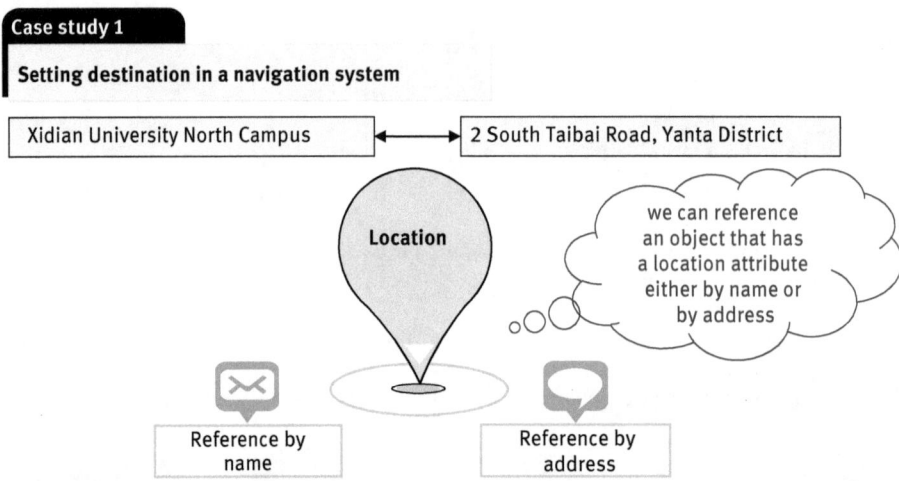

Figure 2.1: Setting destination in a navigation system.

seat is a "reference by address." When assigning homework, a teacher may use the following statements: "questions 6 and 8 of chapter 3" or "question 6 and 8 on page 126." The chapter here is a "reference by name," while the page number is a "reference by address," as shown in Figure 2.2.

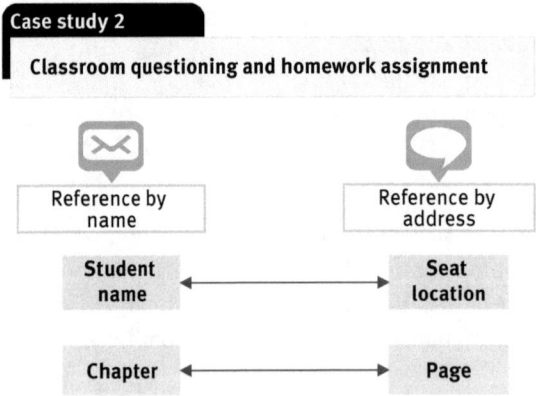

Figure 2.2: Classroom questioning and homework assignment.

We just saw in these examples that we could access a real-life object with location attributes through its name or its address.

Data in programs are also objects with location attributes, so we can use the same methods to access them.

Case study 3 Data reference in programs

As of now, we have been accessing data through referencing variable names. For example, we may use the name x to refer to variable x. Theoretically, we can also access data through their addresses, as long as we have designed a mechanism for it. In fact, we have learned in the introduction of scanf function that we could add a "&" sign in front of a variable to reference it by address. In other words, &x returns the address of variable x, as shown in Figure 2.3.

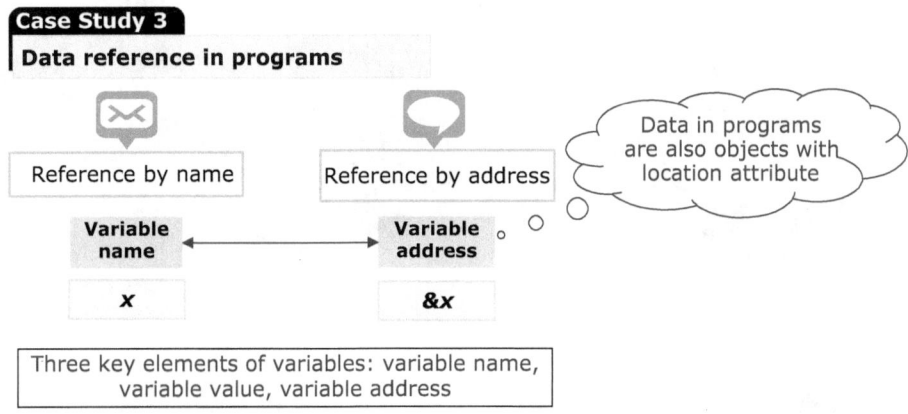

Figure 2.3: Data reference in programs.

With the concept of variable addresses in mind, we can study how computers manage their memory space.

2.1.2 Management of storage space

We shall start from a storage space management problem in practice.

Case study 1 Lockers in kindergarten

To help kids that cannot recognize numbers well remember their lockers, teachers in a kindergarten attached animal stickers to all lockers, as shown in Figure 2.4. Because stickers are more intuitive, the kids are less likely to mistake their lockers.

Similarly, it is intuitive and convenient for programmers to use variables named by meaningful identifiers to operate data.

Figure 2.4: Reference by name in stickers.

Case study 2 Lockers in supermarket
As shown in Figure 2.5, the locker in a supermarket is a large cabinet divided into compartments of the same size, each with a number. A program manages the locker: when

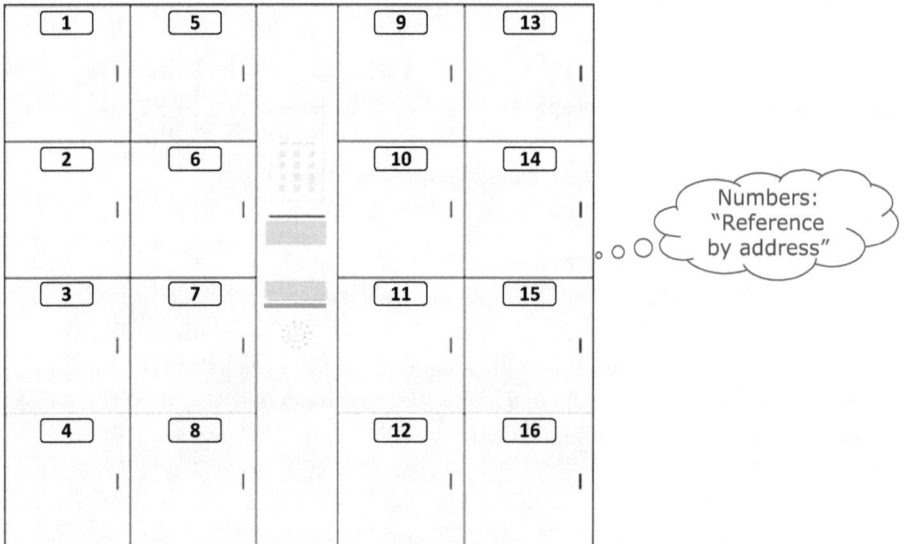

Figure 2.5: Reference by address in lockers.

a customer presses the "Store" button, the system looks for an empty compartment and opens one following specific rule; if no compartment is currently available, "No available compartment" will be displayed.

In this process, the number of the compartment is necessary. If we consider the number of a compartment as its address, then locating compartments through numbers is also "reference by address."

2.1.2.1 Management of computer memory space

After seeing the locker, Mr. Brown thought to himself, "Hey, is not that computer memory?" The memory is where a computer stores programs and data temporarily. As shown in Figure 2.6, we divide the memory into units of the same size to better manage it, which is similar to dividing a locker into compartments. Each unit stores 1 byte (8 bits) of data. They are also called memory units.

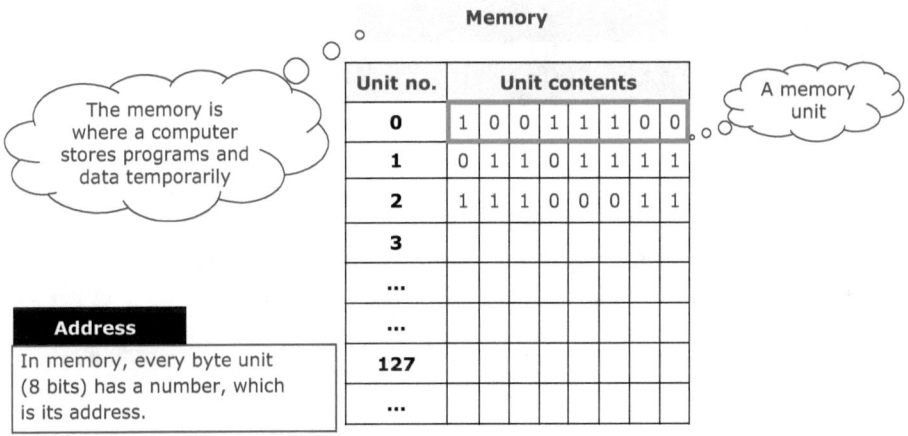

Figure 2.6: Computer memory.

To simplify management, we attach a number to each unit. These numbers are called addresses. A computer can execute memory read and write operations quickly using addresses. The length of a memory unit is 1 byte. Addresses of variables are numbers of memory units allocated by the system.

Knowledge ABC Representation of memory and addresses

Memory is used to temporarily store intermediate computation results of CPU and data that will be exchanged with external memory, such as hard disks. As long as a computer is powered on, the CPU fetches data it needs to the memory and sends the result out after computation is done.

Data are stored in binary form in computers. Addresses are also expressed and processed in binary form. Memory addresses can be expressed in binary, octal, or hexadecimal forms. Assembly languages and high-level languages often use hexadecimal addresses for convenience.

It is also trivial to convert between hexadecimal and binary representations. C uses prefix 0x to represent hexadecimal numbers, while assembly languages and some other high-level languages use suffix H (Hexadecimal).

2.1.2.2 Storage rules of data in memory

Suppose type int has length of 2 bytes in a computer system. "Now I need to store two backpacks," Mr. Brown muttered and defined a variable x of type int. Because int type took up 2 bytes, the variable could not fit into a single memory unit. What would the system do? The answer is to find two consecutive units and allocate to variable x. In other words, the system determines the number of units needed based on data type specified by programmers, looks for consecutive memory units, and allocates them to the variable.

Suppose the system found two consecutive empty units, 2000 and 2001, and allocated it to variable x, as shown in Figure 2.7. Which one was the address of x then?

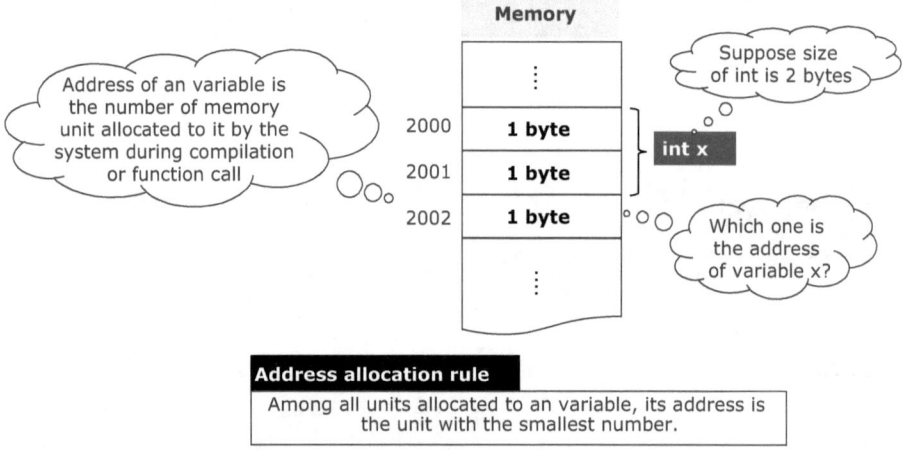

Figure 2.7: Address allocation rule.

In this case, we need the "address allocation rule." The system defines the address of a variable as the unit with the smallest number.

2.1.2.3 Address management in memory

Mr. Brown found it interesting to operate memory directly and mumbled to himself, "I'll be the system administrator this time."

Suppose type int has length of 2 bytes. Professor defined three integer variables i, j, and k. Then he looked for empty units in the memory. In Figure 2.8, empty units are colored in gray. Hence, he could allocate unit 2000 to i, unit 2002 to j, and unit 2004 to k.

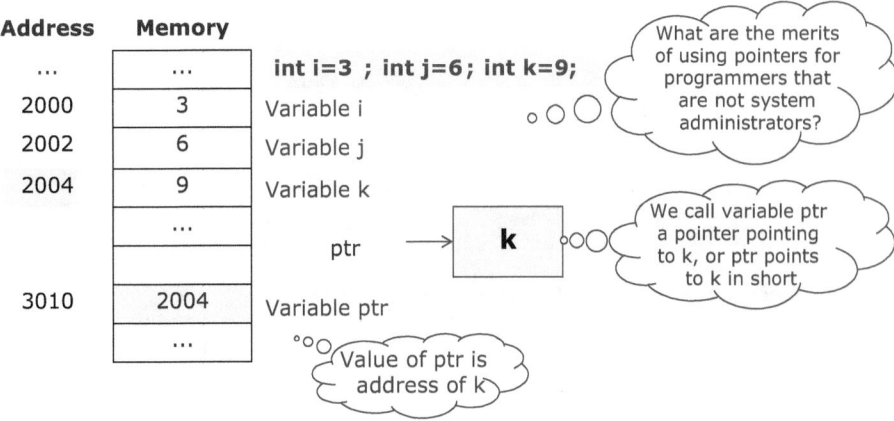

Figure 2.8: Memory usage and management.

Similar to managing a locker, the system needs to record which units are already allocated. This is done using unit addresses as well. As a result, a block of memory is needed to store these addresses. In addition, we should be able to reference it by unit addresses.

How should addresses be recorded?

Professor defined a special variable ptr to store the address of variable k. We often use an arrow to indicate the relation between pointer ptr and variable k. The value of ptr can be changed, so ptr can be used to store other variable addresses as well. In C language, we call variable ptr a pointer pointing to variable k. In short, ptr points to k.

With the help of pointers, system administrators can manage addresses in memory.

Mr. Brown then stopped being an imaginary system administrator, returned to reality by slightly shaking his head, and asked himself another question from the perspective of a computer user: "as programmers, what is the advantage of using address variables?"

It is undoubtedly more intuitive and convenient for programmers to use variables named by identifiers. However, the system has to find the memory unit that corresponds to a variable name when executing the program. This process slows down computation, so the system enables programmers to operate memory units through pointers to enhance execution efficiency, as shown in Figure 2.9. Furthermore, using

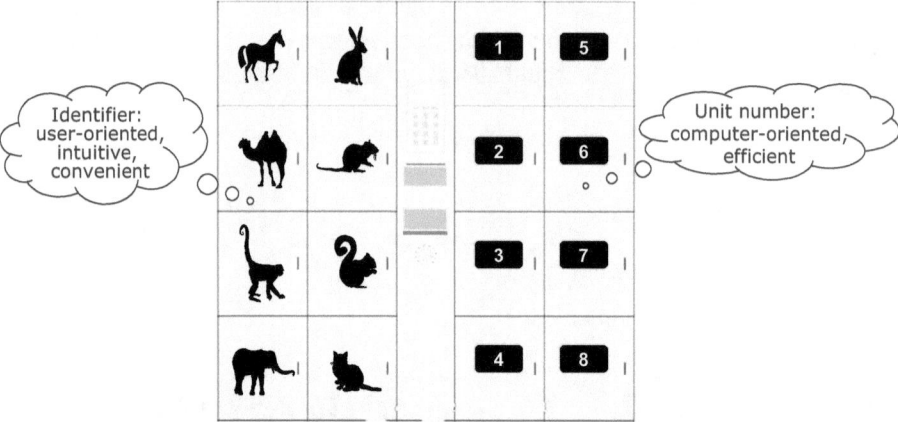

Figure 2.9: Reference by name and reference by address in computers.

pointers help solve problems like batch transfer of data or user space requesting. We will cover related topics in the chapter "Functions."

Having discussed the topics earlier, we shall now proceed to introduce pointers formally.

2.1.3 Definition of pointers

Pointer variables are variables whose values are memory addresses.

2.1.3.1 Comparison of pointer variables and plain variables

Because pointer variables are variables as well, we shall compare them with plain variables and try to find differences between them.

Variables have three key elements: name, content, and address. Names of plain variables are identifiers; contents of plain variables are numerical values; addresses are memory units' numbers. The first element and the third element remain the same for pointer variables, but the contents are different: the value of a pointer variable has to be an address.

Another question related to the nature of pointer value is "what is the type of a pointer?"

For plain variables, C defines variable types as types of their values. However, the definition changes for pointer variables. The language has a special rule for these special variables, as shown in Figure 2.10.

Three key elements of variables		Plain variable	Pointer variable
	Name	Identifier	Identifier
	Value	Number	**Address**
	Address	Memory unit number	Memory unit number

What is the type of a pointer?

Rules

Pointers are special variables. Unlike plain variables, they
- are used to store addresses;
- use types of data stored in the memory units they point to as their own type

Figure 2.10: Similarities and differences between pointer variables and plain variables.

2.1.3.2 Syntax of pointer definitions

Let us look at the syntax of defining a pointer variable. Compared with the syntax of defining a plain variable, the only difference is the * mark in front of the variable name, as shown in Figure 2.11.

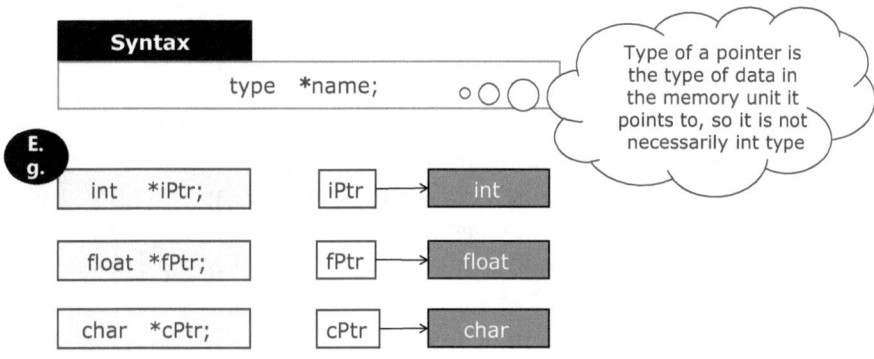

Syntax

type *name;

Type of a pointer is the type of data in the memory unit it points to, so it is not necessarily int type

E.g.

int *iPtr;	iPtr → int
float *fPtr;	fPtr → float
char *cPtr;	cPtr → char

Figure 2.11: Defining a pointer.

C considers the type of a pointer to be the type of data stored in the memory unit it points to, so it is not necessarily an integer. For example, an integer pointer iPtr points to an integer memory unit; a float pointer fPtr points to a floating-point number memory unit; a character pointer works the same way.

Because using pointers requires special rules, it is recommended to name them using "ptr," the abbreviation of "pointer," as a kind reminder.

2.2 Pointer operations

Having learned how to access a pointer, we can now handle pointer data.

2.2.1 Pointer operators

As shown in Figure 2.12, there are two operators related to pointers: the address-of operator "&" and the dereference operator "*". We can access contents stored in memory units by referencing their addresses with the help of these operators.

Operator	Name	Usage
&	Address-of operator	Obtain addresses of plain variables
*	Dereference operator	Obtain data in the memory unit pointed to by the pointer

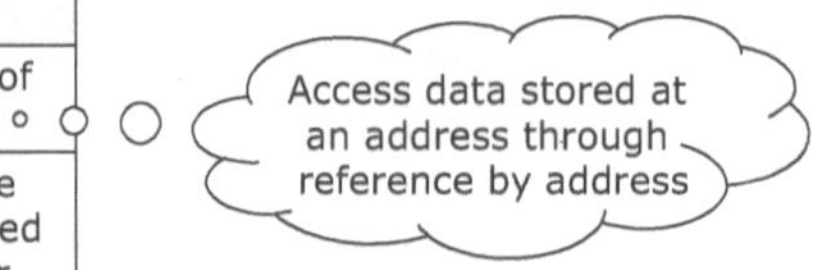

Access data stored at an address through reference by address

Figure 2.12: Pointer operators and their usage.

What operations can we carry out using these operators?

2.2.2 Pointer operations

Unlike operations of plain variables, pointer operations are computations of addresses, so there are only a few types of them, each with certain restrictions. Figure 2.13 shows various pointer operations and their functionalities.

An assignment operation assigns a location to a pointer. An arithmetic operation can move a pointer around and compute the number of elements between two pointers. Relational operations are used to determine the relative position of two pointers.

More pointer operations in arrays will be introduced in Section 2.3.

2.2.3 Basic rules of pointer operations

We shall introduce how to use pointer operators through a simple example.

Example 2.1 Usage of pointers
Suppose we have an integer array x[5], two pointer variables aPtr and bPtr, please write code that completes the following tasks:
- Describing the case illustrated in Figure 2.14.
- Storing the content in the memory unit pointed to by bPtr into the unit pointed to by aPtr.

Operation type	Function	Implementation	Notes
Assignment	Locate a pointer	Assign value to a pointer	The assigned value must be an address of data of the same type as the pointer
Arithmetic	Move a pointer	Add an integer to or subtract one from a pointer	
	Compute number of elements between two pointers	Subtract a pointer from the other	Available for pointers pointing to arrays
Relational	Determine relative position of two pointers	Compare two pointers	

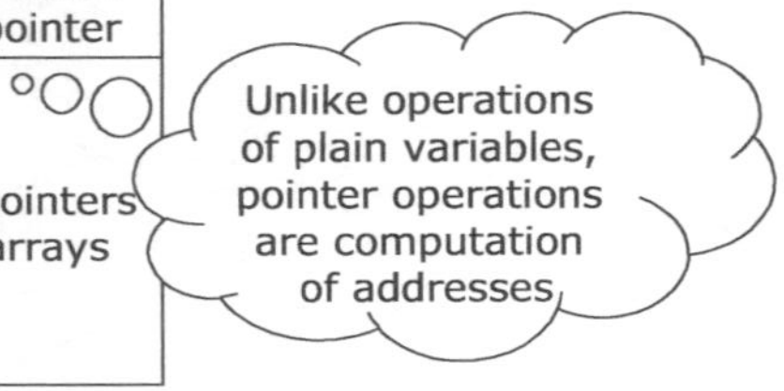

Unlike operations of plain variables, pointer operations are computation of addresses

Figure 2.13: Pointer operations.

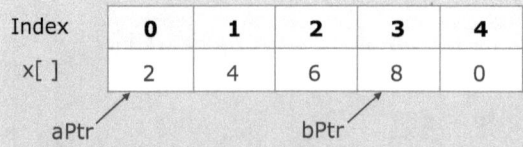

Figure 2.14: Example usage of pointers.

1. Code implementation

Figure 2.15 shows the code implementation, in which we first define the integer array x and initialize it.

On line 2, we define two integer pointers aPtr and bPtr.

On line 3, we make aPtr point to the beginning address of array x, namely the 0-th unit, with statement aPtr = x. By definition, the array name x represents the beginning address, which is the address of element with index 0.

On line 4, we make bPtr point to element with index 3. The & sign is used to obtain the address of x[3].

The first 4 lines complete the first task in the problem description. Now we are going to complete the second.

On line 5, we use *bPtr to fetch value stored in the memory unit pointed to by bPtr. Similarly, the value stored in the memory unit pointed to by aPtr can be acquired by *aPtr. After the assignment, this value becomes 8.

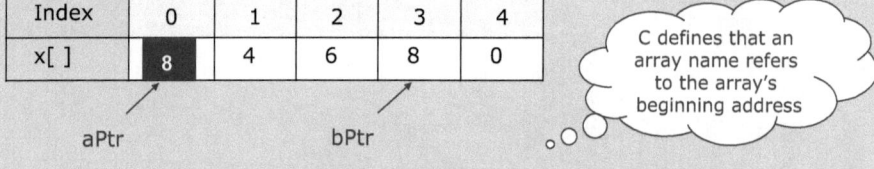

```
01 int x[5]= {2,4,6,8}; //Define and initialize an integer array
02 int *aPtr, *bPtr;     //Define two integer pointers
03 aPtr =x;              // Point aPtr to the beginning address of x
04 bPtr =&x[3];          // Point bPtr to address of x[3]
05 *aPtr =*bPtr;         // Assign value pointed to by bPtr to unit pointed to by aPtr
```

Figure 2.15: Code implementation of the example.

2. Debugging

Using data in the Watch window and the Memory window, we obtain the graph shown in Figure 2.16, which reveals relations between addresses and data.

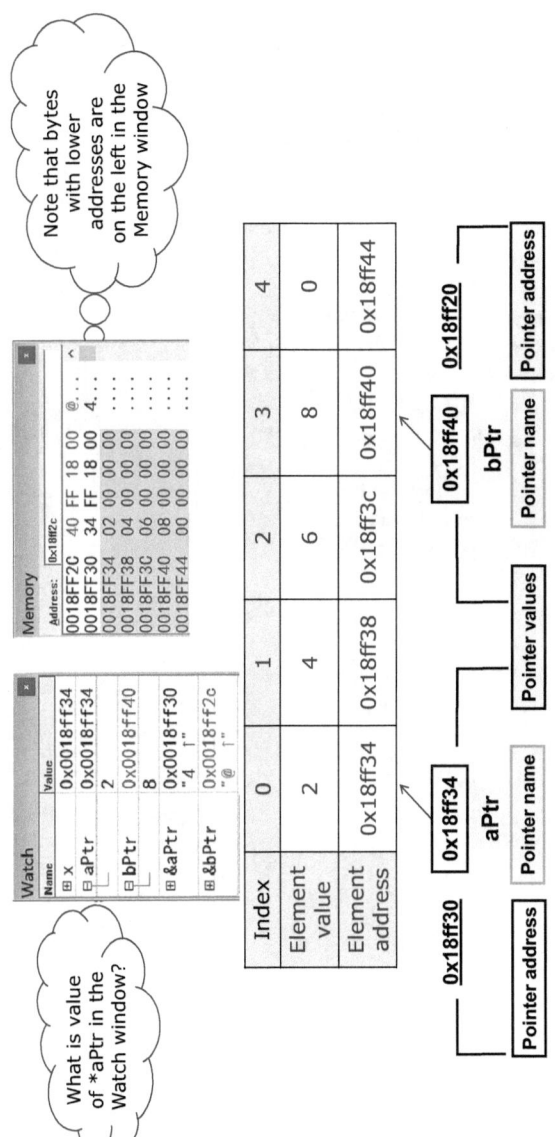

Figure 2.16: Debugging the pointer program.

We shall inspect array x first. In the Watch window, it is clear that the beginning address of x is 0x18ff34. The Memory window shows the addresses and values of elements in x starting from address 0x18ff34. Note that bytes in the Memory window are displayed in the order of their addresses, where bytes in lower addresses are on the left. We can list these addresses and values in a table for further analysis.

It is known that aPtr points to x[0] and bPtr points to x[3]. This can be verified by comparing the value of pointer aPtr and the address of x[0] in the Watch window. We can see that the value of pointer bPtr and the address of x[3] are also identical. Values of aPtr and bPtr are shown in the square above them in the figure. They are both addresses of other variables.

What are the addresses of memory units in which these pointers are stored then?

In the Watch window, &aPtr indicates the address of the memory unit in which aPtr is stored, which is 0x18ff30. Similarly, &bPtr shows the address of bPtr. As of now, we have seen all three key elements of pointer variables, namely variable names, variable addresses, and variable values.

There remains one last question: what is *aPtr then?

We can derive the answer from definitions and verify them in the debugger. In fact, the answer is already given in the next line of aPtr in the Watch window.

3. Exceptions of pointers

If a postman is going to deliver to a new location and heads to the default address without setting a destination in the navigation system, we can well imagine that he will not succeed. Similarly, beginners may make the same mistake when using pointers.

In Figure 2.17, what will happen if we make aPtr point to nowhere by removing the third line in the program?

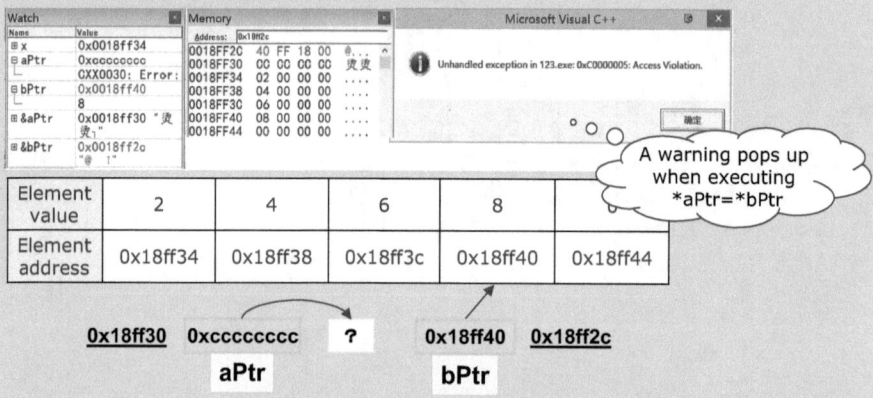

Figure 2.17: Exception of pointers.

After debugging, we can see that everything in the Watch window and the Memory window remains the same, except the value of aPtr being 0xcccccccc. This is because we did not initialize aPtr with the beginning address of array x. The value is determined by the compiler instead of by programmers, so it is unpredictable. Such pointers are often called "wild pointers."

During the execution of line 5, namely *aPtr = *bPtr, a protection mechanism interferes: a warning window pops up, and the program is terminated. This protection mechanism prevents users from writing data to unknown units, so data w be modified unknowingly.

Pointer variables are special variables that need special care. The most common mistakes one may make when using pointers is accessing them without assigning initial values. Figure 2.18 elaborates on this mistake and introduces principles we should follow.

Principles of using pointers

- We should clearly know where our pointers point to
- We should clearly know what data our pointers point to

Programming mistake

It is wrong to assign value to a pointer that is not correctly initialized or points to an unspecified location in memory

Figure 2.18: Principles of using pointer and common mistakes.

These principles are also critical issues in using pointer. Assigning to a pointer that does not point to a certain location has two cons:
(1) It may lead to severe runtime errors, namely logic errors, which affects program execution and may crash the system in the worst case.
(2) Even if the program runs without crashing, the assignment illegally modifies data stored in some memory unit that we do not know. Such errors are extremely hard to find during debugging because we do not know when the modified data are going to be used. If we cannot reproduce such errors, they will become one of the most challenging errors to debug.

2.2.4 Purpose of pointer offsets

2.2.4.1 Introduction

Suppose we have an integer array a, and a pointer aPtr pointing to a[0]. We are asked to output the value of the memory unit aPtr points to, and value of the next unit using aPtr as well.

The previous example showed how to point a pointer to an array element and obtain its value, as shown in Figure 2.19.

To move pointer aPtr forward, we can certainly use reference by address, or to be more specific, aPtr = &a[1]. &a[1] returns the location of the element with index 1 in array a. This is an example of reference by array names. When executing this statement, the system needs to convert &a[1] into memory address corresponding to the element, which may become inconvenient and inefficient when moving the pointer multiple times. Are there alternative methods of moving pointers?

Can we use reference by address by adding an offset to the pointer? If so, it would be easier to move a pointer forward multiple times, as shown in Figure 2.20.

Index	0	1	2	3	4
a[ ]	2	4	6	8	0

aPtr

```
aPtr =a;                      // Point aPtr to the beginning address of a
printf("%d",*aPtr);           // Print data pointed to by aPtr
```

Figure 2.19: Obtain array elements through reference by address.

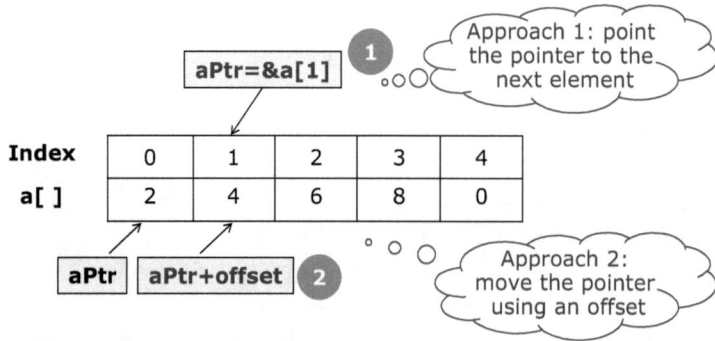

Figure 2.20: Two reference approaches of array elements.

2.2.4.2 Discussion and conclusion

Suppose type int has length of 2 bytes. Furthermore, suppose the memory space of array a is as shown in Figure 2.21 and aPtr points to address 2000.

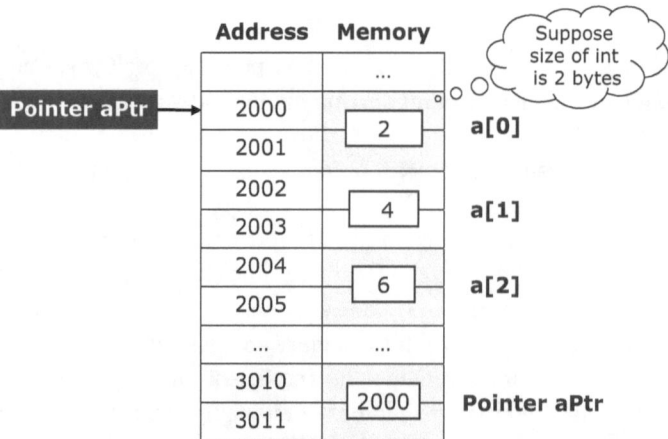

Figure 2.21: Memory and pointers.

Figure 2.22 illustrates the memory layout under the assumption that aPtr points to unit 2001 after adding 1 to it. We will verify this assumption below.

First of all, the assumption is not consistent with the definition of "pointer types." Type of a pointer is the type of the memory unit it points to. If aPtr + 1 points to unit 2001, which belongs to the array element a[0], what is the size of the object it points to then? Second, the value stored in unit 2001 contains half of the information of element a[0].

Hence, it is not reasonable to make aPtr + 1 point to unit 2001. It is natural to infer that we should move the pointer by the length of "pointer type" when adding 1 to it. In this case, it only makes sense if aPtr + 1 points to unit 2002.

Figure 2.23 presents the rule of pointer offsets in C language. Adding an integer to or subtracting an integer from a pointer moves a pointer in memory space. Pointer types determine the distance of such moves.

2.2.4.3 Program verification
We shall verify this rule using the program below:

```
01   #include <stdio.h>
02   int main(void)
03   {
04       int a[5]= {2,4,6,8};        //Define and initialize an integer array
05       int *aPtr;                   //Define an integer pointer
06       aPtr =a;                     //aPtr points to the beginning address of a
07       printf("%d",*aPtr);          //Output value of the memory unit a points to
08       aPtr++;                      //Make aPtr point to the next unit
09       return 0;
10   }
```

On line 6, aPtr points to a[0]. On line 8, we add 1 to aPtr and make it point to a[1].

Figure 2.24 shows the debugging information of this program.

Before adding 1 to aPtr, aPtr points to a[0] and aPtr + 1 points to a[1].

After adding 1 to aPtr, aPtr points to a[1] and aPtr + 1 points to a[2].

2.2.5 Concept of null pointer

2.2.5.1 Meaning of NULL
NULL is a constant defined in header file <stdio.h> with value 0. It is used to represent a null pointer.

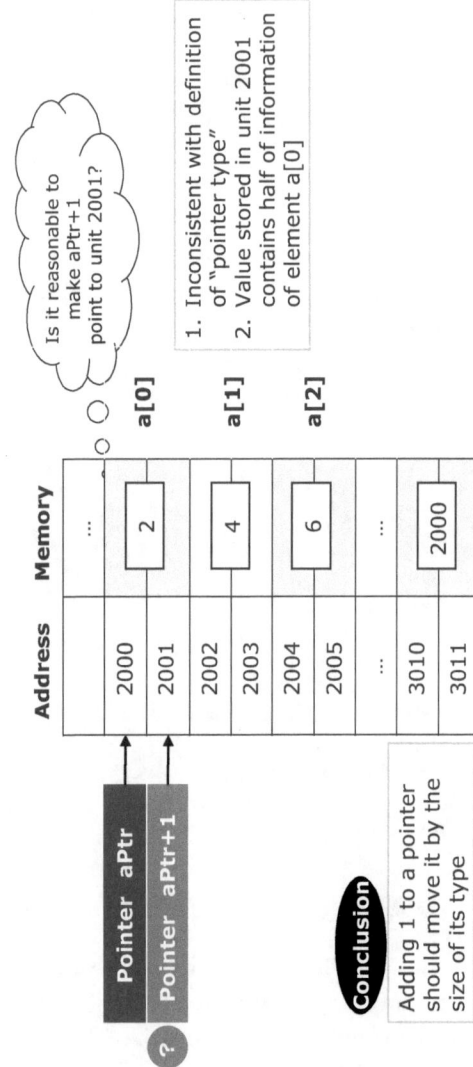

Figure 2.22: Memory and pointer offsets.

Conclusion

The distance a pointer moves past in one shift is the size of its type.

When adding 6 to a float pointer,actual offset is
6*sizeof(float)=24bytes;
When subtracting 7 from a char pointer, actual offset is
7*sizeof(char)=7 bytes;

Figure 2.23: Rule of pointer offsets.

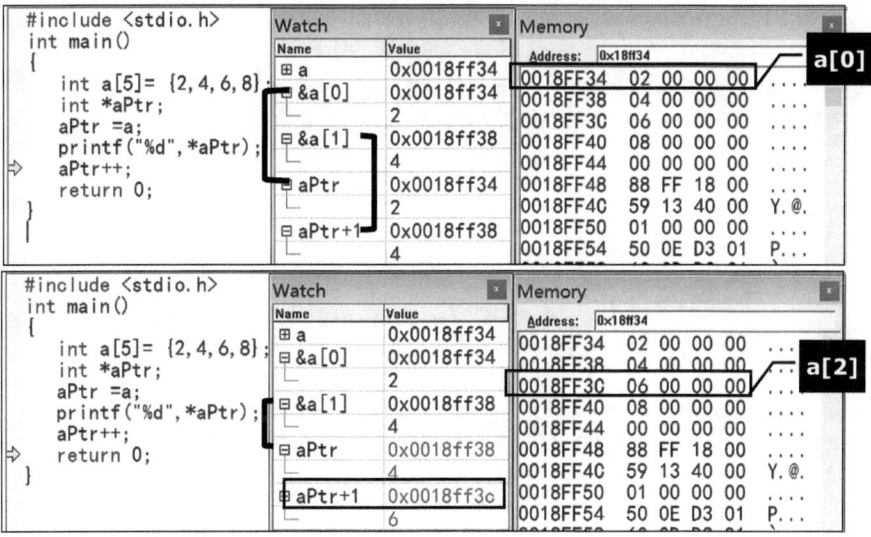

Figure 2.24: Debugging information of pointer offsets program.

2.2.5.2 Null pointer

If we assign NULL to a typed pointer variable, this pointer becomes a null pointer that does not point to any object or function. That is, a null pointer does not point to any memory unit.

A null pointer is not an uninitialized pointer. It is guaranteed that a null pointer does not point to any object, but an uninitialized pointer may point anywhere. Note that a null pointer is not the memory address 0 of a computer.

The purpose of introducing null pointers is that we can return NULL in exception routines so that it is easier to distinguish from a normal address value.

2.3 Pointers and arrays

Pointers are used to reference array elements by address. We shall introduce how to use them in one-dimensional and two-dimensional arrays.

2.3.1 Pointers and one-dimensional arrays

Example 2.2 Computing sum using reference by address
Given a student's grades in 6 classes, please compute the sum using reference by address.

Analysis
We can use the algorithm of the "computing sum" problem in section "one-dimensional arrays operations", but we need to reference array elements by address in this problem instead of by name.

1. Data structure design
We can obtain values of array elements by defining a pointer ptr that points to the array, as shown in Figure 2.25.

Index	0	1	2	3	4	5
score[ ]	80	82	91	68	77	78

ptr

Figure 2.25: Referencing array elements by address.

2. Algorithm description
Figure 2.26 shows the pseudo code. The top-level pseudo code and the first refinement remain the same as before because data reference details are not involved at these two levels. In the second refinement, we start with the initialization of variables and making ptr point to the data array. Then we construct the loop of repeated addition by determining loop control variable, loop execution condition, and offset of ptr.

Top-level pseudo code	First refinement	Second refinement
Scores are stored in array score[6] Compute sum of array elements	The sum is total, scores are stored in score[6]	Initialize score[6], total =0, i=0
		ptr=score;
		while (i<6)
	Add values of elements of score to total repeatedly	total += *ptr;
		i++;
		ptr++;
	Output result	Output total

*ptr fetches value of the unit pointed to by ptr

Figure 2.26: Pseudo code of computing sum algorithm.

```
01   Compute sum of array elements
02 #include <stdio.h>
03 #define SIZE 10
04
05   main(void)
06 {
07     score[ SIZE ] = {98,92,89,95,90,96,94,92,90,97};
08     i;                    // Counter
09     total = 0;            // Sum
10
11     ( i = 0; i < SIZE;  i++ )
12     {
13         total +=score[ i ];  // Compute sum of score elements
14     }
15     printf( "The total score is %d\n", total );
16        0;
17 }                          Reference by name
```

```
01 //Compute sum of array elements
02 #include <stdio.h>
03 #define SIZE 10
04
05 int main(void)
06 {
07     int score[ SIZE ] = {98,92,89,95,90,96,94,92,90,97};
08     int i, *ptr=score; //ptr points to the array
09     int total = 0;        // Sum
10
11     for ( i = 0; i < SIZE;  i++ , ptr++)
12     {
13         total +=*ptr;   //Compute sum of score elements
14     }
15     printf( "The total score is %d\n", total );
16     return 0;
17 }                          Reference by address
```

The total score is 933

```
int *ptr=score;
is equivalent to
int *ptr;
ptr=score;
```

Figure 2.27: Code implementation of computing sum problem.

3. Code implementation

It is trivial to obtain actual code starting from pseudo code in the second refinement. Figure 2.27 lists programs of reference by name and by address together for readers to compare.

On line 8, pointer ptr is defined. Note that the statement int *ptr = score defines the pointer and assign a value to it.

On line 11, ptr should be increased as well.

On line 13, the element pointed to by ptr is added to the sum.

Example 2.3 Pointer pointing to constant string

Compare a character array with a pointer pointing to a string.

Analysis

The test program is as follows:

```
1  int main(void)
2  {
3      char a[]="dinar##";
4      char *b="dollar##";
5
6      a[6]=':';
7      b[5]=':';
8      return 0;
9  }
```

If we run the program, an "Access Violation" warning will pop up. Debugging shows that the error occurs when executing line 7. That is, we cannot write to the string pointed to by pointer b. This is because a constant string is stored in the constant segment in memory, which cannot be altered during execution; on the other hand, assigning a constant string to an array essentially puts the string into the variable segment, which can be modified. Readers can refer to chapter "Functions" for more details on memory layout.

Conclusion Pointers and constant strings

We cannot write to the memory segment in which constant strings are stored.

Example 2.4 Program analysis

Analyze the following program and list values of memory units pointed to by pointer aPtr and bPtr in each iteration:

```
1   int main(void)
2   {
3       int a[10], b[10];
4       int *aPtr, *bPtr, i;
5       aPtr=a; bPtr=b;
6       for ( i=0; i<6; i++, aPtr++, bPtr++)
7       {
8           *aPtr=i;
9           *bPtr=2*i;
10      printf("%d\t%d\n", *aPtr,* bPtr);
```

```
11  }
12  aPtr=&a[1];      //Step 1
13  bPtr =&b[1];     //Step 2
14  for (i=0; i<5; i++)
15  {
16   *aPtr +=i;       //Step 3
17   *bPtr *=i;       //Step 4
18   printf("%d\t%d\n", *aPtr++,* bPtr ++);
19  }      //*aPtr++ fetches value first, and then adds 1 to aPtr
20  return 0;
21  }
```

Analysis

Figure 2.28 shows values of arrays a and b after the for loop on line 6 terminates.

a	0	1	2	3	4	5
b	0	2	4	6	8	10

Figure 2.28: Values of arrays a and b.

Since we have aPtr = &a[1] and bPtr = &b[1] in step 1 and 2, values of *aPtr and *bPtr should be 1 and 2, respectively. Figure 2.29 shows these two values in each iteration of the for loop on line 14, starting from i = 0.

i	0	1	2	3	4
*aPtr in step ①	1	2	3	4	5
*aPtr in step ③	1	3	5	7	9
*bPtr in step ②	2	4	6	8	10
*bPtr in step ④	0	4	12	24	40

Figure 2.29: Data analysis table.

Example 2.5 Program analysis

Analyze the following program, figure out all objects pPtr and sPtr point to during execution and the program result:

```
1  int main(void)
2  {
3    char a[2][5]={"abc","defg"};
4    char *pPtr=a[0],*sPtr=a[1];
5    while (*pPtr) pPtr++;
6    while (*sPtr) *pPtr++=*sPtr++;
7    printf("%s%s\n",a[0],a[1]);
8    return 0;
9  }
```

Analysis

Figure 2.30 illustrates the case where pointers pPtr and sPtr point to array a.

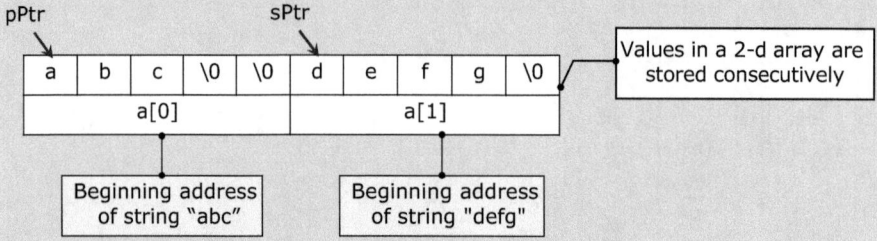

Figure 2.30: Pointers pointing to array a.

(1) Figure 2.31 illustrates the case after line 5 is executed.

Figure 2.31: Program analysis 1.

(2) sPtr points to "d" at first. In the loop "while (*sPtr) *pPtr++ = *sPtr++" on line 6, we repeatedly assign value pointed to by sPtr to the unit pointed to by pPtr. In the first iteration, the value is updated to "d," then both pointers move to the next element, as shown in Figure 2.32.

Figure 2.32: Program analysis 2.

(3) Now sPtr points to "e." After another iteration, the value of the unit pointed by pPtr is updated to "e," as shown in Figure 2.33.

Figure 2.33: Program analysis 3.

(4) Now sPtr points to "f." After another iteration, the value of the unit pointed by pPtr is updated to "f," as shown in Figure 2.34.

a	b	c	\0	\0	d	e	f	g	\0
			d	e	f				

pPtr → (above \0, fifth cell); sPtr → (above f, eighth cell)

a[0]   a[1]

Figure 2.34: Program analysis 4.

(5) Now sPtr points to "g." After another iteration, the value of the unit pointed by pPtr is updated to "g," as shown in Figure 2.35.

a	b	c	\0	\0	d	e	f	g	\0
			d	e	f	g			

pPtr → (above d, sixth cell); sPtr → (above g, ninth cell)

a[0]   a[1]

Figure 2.35: Program analysis 5.

(6) Now sPtr points to "\0." Because the loop condition is not met, the loop terminates, as shown in Figure 2.36.

a	b	c	\0	\0	d	e	f	g	\0
			d	e	f	g			

pPtr → (above f, seventh cell); sPtr → (above \0, tenth cell)

a[0]   a[1]

Figure 2.36: Program analysis 6.

(7) On line 7, %s format specifier prints character starting from the given address and stops upon reaching "\0." Hence, the string starting at address a[0] is abcdefgfg and the string starting from address a[1] is fgfg. As a result, the final output is abcdefgfgfgfg.

2.3.2 Pointers and two-dimensional arrays

Example 2.6 Computing total grade for multiple students
Suppose we have four students and their grades in six courses, as shown in Figure 2.37. Please compute the total grade for each of them using referencing by address.

ID	Course 1	Course2	Course3	Course4	Course5	Course6	Total
1001	80	82	91	68	77	78	
1002	78	83	82	72	80	66	
1003	73	50	62	60	75	72	
1004	82	87	89	79	81	92	

We can repeat the algorithm for one student 4 times

Figure 2.37: Grades of students.

Analysis
We can simply apply the same algorithm four times to complete the task.

1. Data analysis
Let us analyze the data to be processed first. Fetching address for each element in a row can be done in the same way as in one-dimensional arrays. We will use a pointer ptr and update it to the beginning address of the next row after processing one row.

To obtain the address of the next row, we can certainly reference the one-dimensional row by score[1]. However, this is a reference by name instead of by address. Can we reference the row by address then? In other words, can we use another pointer sPtr as a row pointer for the two-dimensional array, as shown in Figure 2.38?

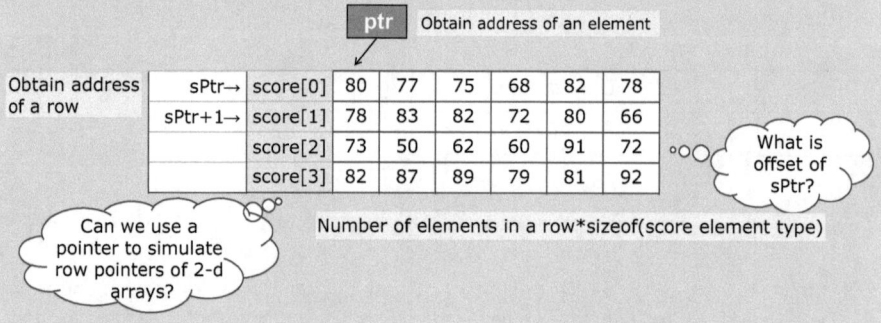

Figure 2.38: Row pointer of two-dimensional array.

The answer is affirmative because we can treat score[0] to score[3] as array elements as well. The next question is what should be the offset of sPtr?

According to the definition of pointer offset, it should be the number of elements in a row multiplied by the size of elements type.

There is a special term in C for pointers pointing to row addresses of a two-dimensional array. As shown in Figure 2.39, these pointers are called "pointer to arrays." This is a confusing term, so parentheses are added to pointer names to distinguish them from plain pointers and arrays.

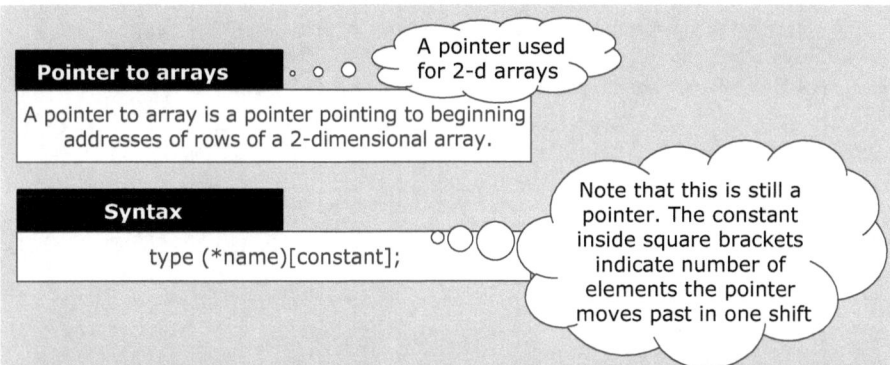

Figure 2.39: Pointer to arrays.

It is worth noting that this is still a pointer, even though there is a constant wrapped by square brackets after the pointer name. In fact, the constant indicates the number of elements the pointer moves past in one shift.

For this problem, we can define a pointer to array as shown in Figure 2.40.

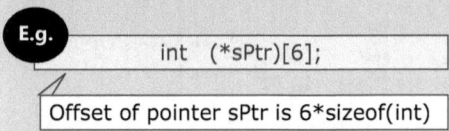

Figure 2.40: Example of pointer to array.

Now we can obtain information of the two-dimensional array through reference by address, and find out relation between ptr and sPtr, as shown in Figure 2.41.

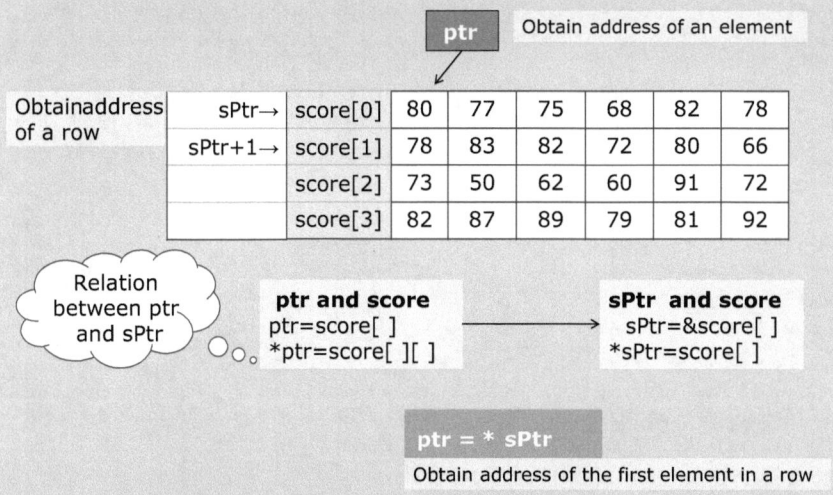

Figure 2.41: Relation between row pointers and element pointers.

The object in the unit pointed to by ptr is an element of array score. The object in the unit pointed to by sPtr is address of a one-dimensional row of array score. Hence, we can write statement ptr = *sPtr, where *sPtr represents the beginning address of a row.

2. Code implementation

```
01  #include <stdio.h>
02  #define N  4 //Number of rows
03  #define M  6 //Number of columns
04  int main(void) {
05
06     int score[N][M]=
07     {
08          {80,77,75,68,82,78},
09          {78,83,82,72,80,66},
10          {73,50,62,60,91,72},
11          {82,87,89,79,81,92}
12     };
13     int i,j;
14     int total;          //Total grade
15     int *ptr;           //Row pointer
16     int (*sPtr)[M];     //A pointer to array, offset is M int
17     sPtr=&score[0];     //Make sPtr point to the first row
18
19     for (i=0; i<N; i++, sPtr++)
20     {
21        total = 0;
22        ptr=*sPtr;              //Make ptr point to the beginning address of the row
23        for ( j= 0;  j< M;  j++, ptr++)
24        {
25           total +=*ptr;    //*ptr=score[ ][ ]
26        }
27        printf( "Total grade of student %d is %d\n", i+1,total );
28     }
29     return 0;
30  }
```

Program result:
Total grade of student 1 is 460
Total grade of student 2 is 461
Total grade of student 3 is 408
Total grade of student 4 is 510

Note: On line 16, we define a pointer to array sPtr with offset being M int, where M is 6.
On line 17, we make sPtr point to the first row of the array.
On line 22, ptr is set to point to the first element in a row.
In the for loop between line 23 and 26, we use pointer ptr to retrieve elements in array score and add them to the sum.

In each iteration, ptr is increased by 1 in the loop increment part. That is, it moves to the next element of the row.

In loop increment part of the for loop on line 19, sPtr is increased by 1. That is, it moves to the next row.

3. Debugging

We can inspect the memory layout in the Watch window and the Memory window. Figure 2.42 shows the beginning addresses of each row of array score.

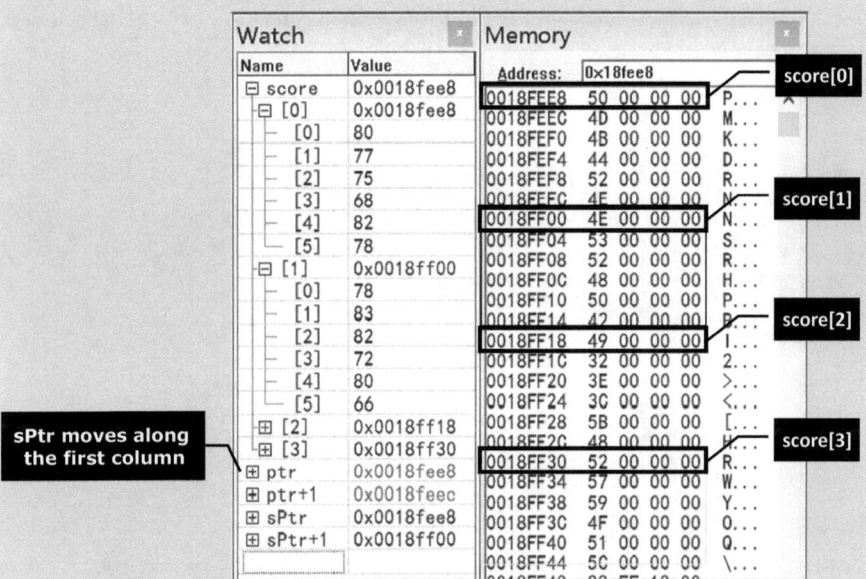

Figure 2.42: Referencing elements of a two-dimensional array.

score[0] is 0x18fee8, which corresponds to the first gray block in the Memory window.

score[1] is 0x18ff00, which corresponds to the first white block.

Similarly, we have score[2] being 0x18ff18 and score[3] being 0x18ff30.

It is clear from the figure how ptr moves along each row. The value of ptr in the Watch window is the address of the 0th element in the score[0] block in the Memory window, which is 0x18fee8. The value of ptr + 1 is the address of the 1st element in score[0] block, which is 0x18feec.

sPtr moves along the first column of score. We have sPtr = score[0] and sPtr + 1 = score[1] at first.

Values are displayed as hexadecimal numbers in the Memory window. The value of score[0][2] is 4B, which corresponds to its decimal form 75 in the Watch window.

It can be derived from the Memory window that int type takes up 4 bytes in the system in which this program is executed. The address of the last element in the 0th row is 0x18fefc. After shifted by 4 bytes, it becomes 0x18ff00, which is precisely the address of the first element in the first row. This shows that the addresses of these two elements are consecutive. Similarly, one can inspect the addresses of the first and last elements of other rows and conclude that rows of a two-dimensional array are stored consecutively.

Once again, we notice that elements of multidimensional arrays are stored consecutively, which is a general rule of array storage.

2.4 Pointers and multiple strings

Example 2.7 Finding largest string
Find the largest string (in alphabetical order) in the following family names. Please implement using reference by address:
Zhao, Zhou, Zhang, Zhan, Zheng

Analysis

1. Data structure analysis
Each family name is a string, so we can store multiple family names in a two-dimensional character array, as shown in Figure 2.43.

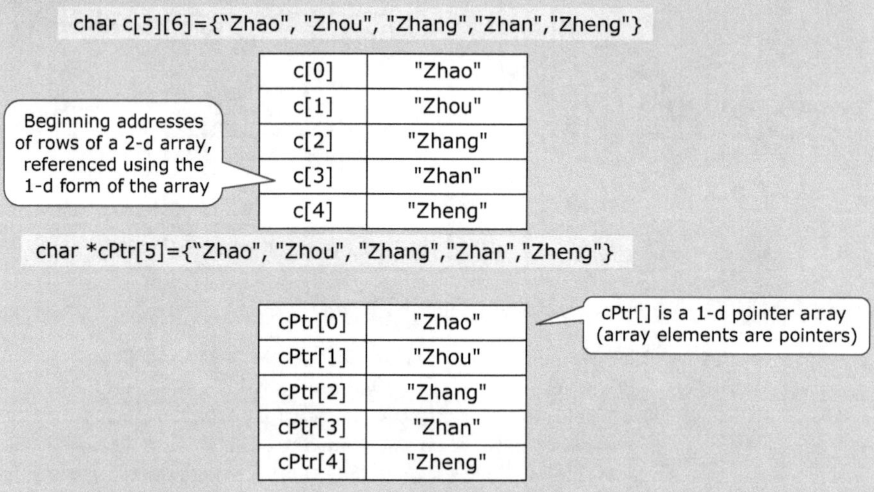

char c[5][6]={"Zhao", "Zhou", "Zhang","Zhan","Zheng"}

c[0]	"Zhao"
c[1]	"Zhou"
c[2]	"Zhang"
c[3]	"Zhan"
c[4]	"Zheng"

Beginning addresses of rows of a 2-d array, referenced using the 1-d form of the array

char *cPtr[5]={"Zhao", "Zhou", "Zhang","Zhan","Zheng"}

cPtr[0]	"Zhao"
cPtr[1]	"Zhou"
cPtr[2]	"Zhang"
cPtr[3]	"Zhan"
cPtr[4]	"Zheng"

cPtr[] is a 1-d pointer array (array elements are pointers)

Figure 2.43: Two storage structures of multiple strings.

Because there are five names, we need to define a character array c with five rows. The longest name has five characters, so we need six columns to store it and the termination mark.

The address of each row of a two-dimensional array can be referenced in a one-dimensional format. c[0] to c[4] can be treated as elements of a special array whose elements are pointers.

Based on discussion earlier, we may define a pointer array cPtr[]. It is a one-dimensional array of pointers with five elements, each of which is the address of a string.

2. Algorithm description
The pseudo code is shown in Figure 2.44.

Top-level pseudo code	First refinement	Second refinement
Find the largest among multiple strings	Store M strings into *cPtr[M]	char *cPtr[M] , char str[6]
	Use the first string in the array as comparison basis str	Use cPtr[0] as comparison basis, and copy it into str
	Compare each string with str and put the larger into str	i=1;
		while i< M
		if str<cPtr[i]
		Copy c[i] into str
		i++;
Output result	Output result	Output str

Figure 2.44: Finding largest string algorithm.

3. Code implementation

```
01   #include <stdio.h>
02   #include <string.h>
03   #define M 5        //Number of strings
04   #define N 5        //Longest string length + 1
05
06   int main(void)
07   {
08       char *cPtr[M]= {"Zhao","Zhou","Zhang","Zhan","Zheng"};
09       char str[N];
10       int  i;
11       //Use strcpy to copy cPtr[0] into str, beware of out-of-bound error
12       strcpy(str, cPtr[0]);
13       for (i=1; i<M; i++)
14       {
15          if (strcmp(str, cPtr[i])< 0)        //If str is less than cPtr[i]
16          {
17             strcpy(str, cPtr[i]);            //Then copy cPtr[i] into str
18          }
19       }
20       printf("The largest string is: %s\n", str);
21       return 0;
22   }
```

Program result:
    The largest string is: Zhou

Note: On line 8, pointer array cPtr is defined and initialized with 5 strings.

   On line 9, a one-dimensional character array is defined. Note that its length is the number of characters in the longest string plus one.

   The watch window shows the elements of the pointer array, as shown in Figure 2.45. Values of these elements are the beginning addresses of strings.

Watch		
Name	**Value**	
⊟ cPtr	0x0018ff34	
⊞ [0]	0x00420f9c	"Zhao"
⊞ [1]	0x00420f94	"Zhou"
⊞ [2]	0x00420f8c	"Zhang"
⊞ [3]	0x00420034	"Zhan"
⊞ [4]	0x0042002c	"Zheng"

Figure 2.45: Inspection of pointer array.

On line 12, we use strcpy function to copy cPtr[0] into array str.
On line 15, we compare strings stored in str and cPtr[i].
On line 17, we copy the larger one into str.

2.4.1 One-dimensional pointer array and pointer to pointer

In the earlier example, if we wish to use a pointer to point to elements in array cPtr, what does this pointer that points to pointer array look like?

Let the pointer pointing to pointer array cPtr be cPtrPtr, then its relation with elements in array cPtr should be as shown in Figure 2.46. In this case, we need a new kind of pointers.

1-dimensional pointer array
char *cPtr[5]={"Zhao", "Zhou", "Zhang","Zhan","Zheng"} ;
//Array elements are pointers

We want a pointer pointing to a pointer array, what does it look like?

cPtrPtr→	cPtr[0]	"Zhao"
cPtrPtr+1→	cPtr[1]	"Zhou"
	cPtr[2]	"Zhang"
	cPtr[3]	"Zhan"
	cPtr[4]	"Zheng"

Figure 2.46: One-dimensional pointer array.

cPtrPtr points to cPtr, which is a pointer, so cPtrPtr is a pointer to pointer. We usually call such pointers double pointers. Figure 2.47 shows how to define a double pointer and how to assign a value to it. A double pointer requires two asterisks in front of the pointer name.

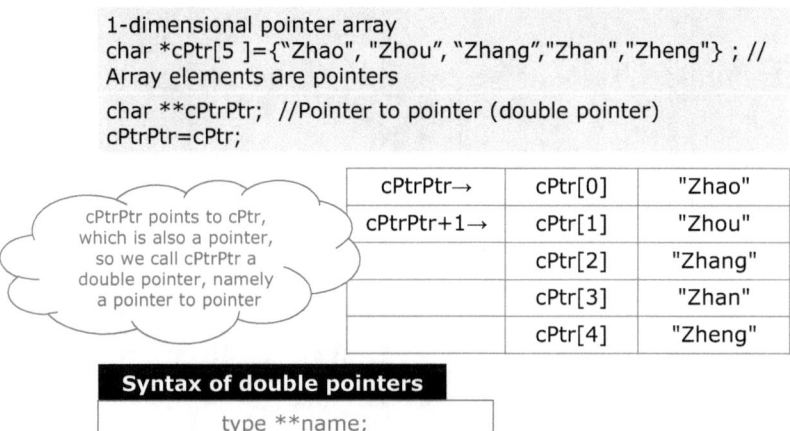

1-dimensional pointer array
char *cPtr[5]={"Zhao", "Zhou", "Zhang","Zhan","Zheng"} ; //
Array elements are pointers

char **cPtrPtr; //Pointer to pointer (double pointer)
cPtrPtr=cPtr;

cPtrPtr→	cPtr[0]	"Zhao"
cPtrPtr+1→	cPtr[1]	"Zhou"
	cPtr[2]	"Zhang"
	cPtr[3]	"Zhan"
	cPtr[4]	"Zheng"

cPtrPtr points to cPtr, which is also a pointer, so we call cPtrPtr a double pointer, namely a pointer to pointer

Syntax of double pointers

type **name;

Figure 2.47: Definition of pointer to pointer.

To inspect a double pointer in debugger, we can use the three statements shown in Figure 2.48.

1-dimensional pointer array
char *cPtr[5]={"Zhao", "Zhou", "Zhang","Zhan","Zheng"} ;
// Array elements are pointers
char **cPtrPtr; //Pointer to pointer (double pointer)
cPtrPtr=cPtr;

Watch			Memory			
Name	Value		Address:	0x18ff34		
⊟ cPtr	0x0018ff34		0018FF34	94 0F 42 00	. . B.	^
⊞ [0]	0x00420f94 "Zhao"		0018FF38	8C 0F 42 00	. . B.	
⊞ [1]	0x00420f8c "Zhou"		0018FF3C	34 00 42 00	4. B.	
⊞ [2]	0x00420034 "Zhang"		0018FF40	2C 00 42 00	, . B.	
⊞ [3]	0x0042002c "Zhan"		0018FF44	1C 00 42 00	. . B.	
⊞ [4]	0x0042001c "Zheng"		0018FF48	88 FF 18 00	. . . .	
⊟ cPtrPtr	0x0018ff34		0018FF4C	59 13 40 00	Y. @.	
⊞	0x00420f94 "Zhao"		0018FF50	01 00 00 00	. . . .	
⊟ cPtrPtr+1	0x0018ff38		0018FF54	50 0E C5 01	P. . .	
⊞	0x00420f8c "Zhou"		0018FF58	60 0D C5 01	`. . .	
⊞ &cPtr[0]	0x0018ff34		0018FF5C	70 12 40 00	p. @.	
⊞ &cPtr[1]	0x0018ff38		0018FF60	91 91 48 77	慨Hw	
			0018FF64	00 F0 FD 7F	虫	

Figure 2.48: Inspection of the pointer array.

As shown in the Watch window, elements of cPtr are beginning addresses of strings. Addresses of these elements are presented in the Memory window. For example, element cPtr[0] has value 0x420f94, which is the beginning address of the string "Zhao." The address of cPtr[0] is 0x18ff34, as shown in the Memory window. The double cPtrPtr points to elements of cPtr. By adding 1 to it, we move it to the next element.

Readers can reimplement the largest string algorithm using double pointers.

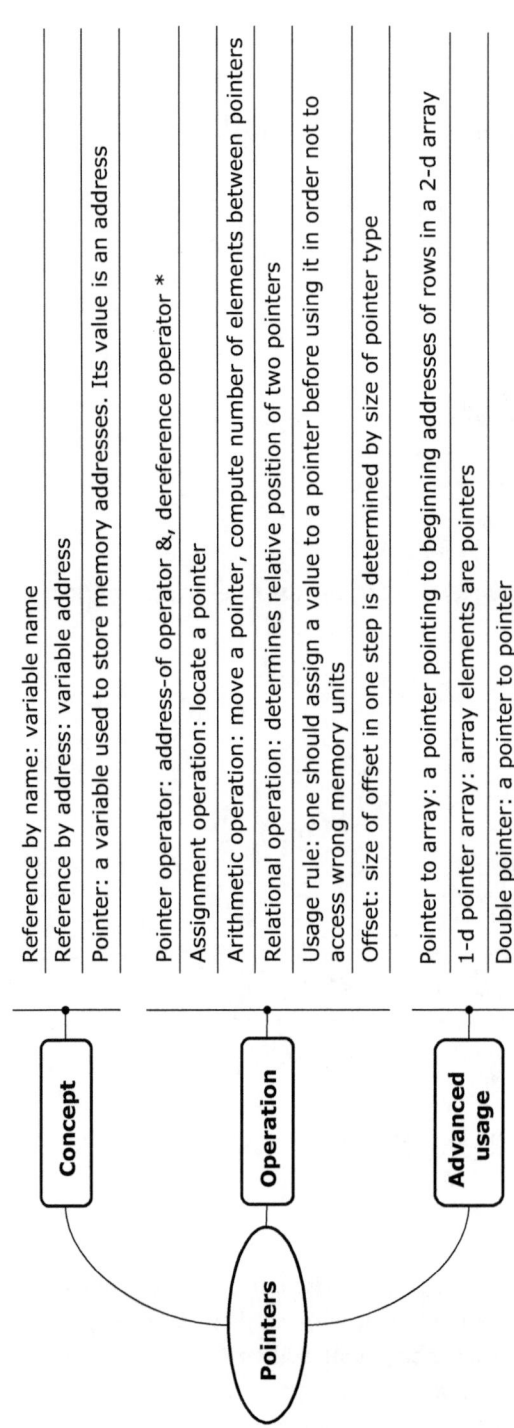

Figure 2.49: Relations between concepts related to pointers.

2.5 Summary

The main contents of this chapter and the relations between them are shown in Figure 2.49.

Objects with location attribute can be referenced either by name or by address,
Which are both key elements of variables,
Variable names are usually meaningful, so it is more intuitive and convenient to use them,
Reference by address operates on memory directly, so it is more efficient.

Data are stored in cells inside memory.
Each cell has a number, which is used as their addresses,
Values of pointer variables are addresses,
And their types indicate types of data stored in the corresponding cell,
To access the data, we need to determine the address first.
Pointer operations are limited to comparison and moving.
The size of step must be determined before moving a pointer.

2.6 Exercises

2.6.1 Multiple-choice questions

1. [Null pointer]
    Which of the following is the output of this program? ( )

    ```
    # include <stdio.h>
    int main(void)
    {
        printf("%d\n",NULL);
        return 0;
    }
    ```

    A) We do not know   B) 0   C) –1   D) 1

2. [Concept of pointers]
    Which of the following statements is correct about addresses and pointers? ( )
    A) We can assign a pointer of one type to a pointer of another type through forced-type conversion.
    B) We can compute the address of a constant and assign it to a pointer of the same type.

C) We can compute the address of an expression and assign it to a pointer of the same type.

D) We can compute the address of a pointer and assign it to a pointer of the same type.

3. [Pointer assignment]

Suppose x is an integer variable and pb is an integer pointer. Which of the following statements is correct? ( )

A) pb = &x   B) pb = x   C) *pb = &x   D) *pb = *x

4. [Pointer exception]

Suppose we have the following definitions: int x=2, *p=&x; float y=3.0; char z='c';.

Which of the following operations is unsafe? ( )

A) p++   B) x++   C) y++   D) z++

5. [Read into address]

Suppose we have declarations double *p, a. Which of the following statements can correctly read input? ( )

A) *p = &a; scanf("%lf",p)   B) p = (double*)malloc(8); scanf("%f",p)

C) p = &a; scanf("%lf",a)   D) p = &a; scanf("%lf",p)

6. [Pointer offset]

What is value of y after executing the following program? ( )

```
int a[]={2,4,6,8,10};
int y=1,x,*p;
p=&a[1];
for(x=0;x<3;x++) y += * (p + x);
printf("%d\n",y);
```

A) 17   B) 18   C) 19   D) 20

7. [Operations on string]

Which of the following statements is a correct string assignment statement? ( )

A) char s[5] = {"ABCDE"}   B) char s[5] = {'A', 'B', 'C', 'D', 'E'}

C) char *s; s = "ABCDEF"   D) char *s; scanf("%s", s)

8. [Forced-type conversion of pointer]

Which of the following options points pointer p to a dynamic memory unit of an integer variable? ()

int *p;

p = _____ malloc(sizeof(int));

A) int   B) int *   C) (*int)   D) (int *)

Note: malloc is the library function for dynamic memory allocation

9. [Row pointer of two dimensional array]
   Suppose we have: int w[3][4] = {{0,1},{2,4},{5,8}}; int(*p)[4] = w;
   Which of the following expressions evaluates to 4? ( )
   A) *w[1]+1   B) p++,*(p+1)   C) w[2][2]   D) p[1][1]

2.6.2 Fill in the tables

1. [Memory unit address]
   Figure out values of variables in Figure 2.50 after executing the following program. Suppose address of variable a is 0x003FFCA4 and address of variable b is
   0x003FFCA8.

```
int main(void)
{
    int a, b;
    int *p1, *p2;
    p1 = &a;
    p2 = &b;
    a = 50;
    b = 20;
    a = *p1 - *p2;
    return 0;
}
```

Variable	a	b	p1	p2	*p1	*p2
Value						

Figure 2.50: Pointers: fill in the tables, question 1.

2. [Pointer operations]
   Fill in the table in Figure 2.51 based on the following program:

```
int a, b, k=4, m=6, *p1=&k, *p2=&m;
a=(p1==&m);
b=(*p1) / (*p2)+7;
```

Expression	Result
p1==&m	
*p1	
*p2	
(*p1)(*p2)	
a	
b	

Figure 2.51: Pointers: fill in the tables, question 2.

3. [Pointer to one-dimensional character array]
   Fill in the table in Figure 2.52 based on the following program:

```
int main(void)
{
    int i, s=0, t[]={1,2,3,4,5,6,7,8,9};
    int *p=t;
    for(i=0;i<9;i+=2)
    {
        s+=*(p+i);
    }
    printf("%d\n",s);
    return 0;
}
```

i	0	2	4	6	8	10
p	&t[0]					End of loop
s	1					

Figure 2.52: Pointers: fill in the tables, question 3.

4. [Pointer to character array]
   Fill in the table in Figure 2.53 based on the following program. Suppose keyboard input is "abcde"
   #include "ctype.h"

```
int main(void)
{
    char str[81],*sptr;
    gets(str);
    sptr=str;
    while(*sptr)
    {
```

```
        putchar(*sptr+1);
        sptr++;
    }
    return 0;
}
```

Input str	"abcde"				
Number of iterations	1	2	3	4	5
*sptr	'a'				
putchar(*sptr+1)	'b'				
Functionality:					

Figure 2.53: Pointers: fill in the tables, question 4.

2.6.3 Programming exercises

1. Suppose we have an integer array with 10 elements. Show the output of its elements using the following methods: through array index, through array name, and through a pointer.
2. Write a program that reads n number from keyboard input and outputs them in the reversed order of input. Your implementation should use pointers.
3. Please write a program, in which you define a one-dimensional integer array num[20], read an integer n (n ≤ 20), and an integer sequence (of n numbers) from keyboard input, find the maximum and the minimum in the sequence and swap them.
4. Write a function that stores input characters backwards. Input of characters and output of the reversed characters should be done in the main function.
5. Write a program that connects two strings without using strcat function.
6. Please write a program that handles character input in the following way: if the input is a lowercase letter, the program should output its uppercase counterpart; if the input is uppercase, the program should output its lowercase counterpart; other characters should be output as such.
7. A palindromic number is a nonnegative integer that remains the same when its digits are reversed, for example, 12321. Please write a program that determines whether the input integer is palindromic. If so, the program should output the sum of its digits; otherwise the program should output "no."
8. Please write a program to encrypt string "China" using Caesar code with a right shift of 4. For example, the fourth letter after "A" is "E," so ciphertext of "A" is "E." As a result, ciphertext of "China" should be "Glmre."

3 Composite data

Main contents
- Introduction of construction of structures through comparison of structures and arrays
- Analysis of the nature of structure types through comparison of structure types and basic types
- Summarization of usage of structures through comparison of structure members and plain variables
- Program reading practices
- Practice of top-down stepwise refinement algorithm design
- Storage characteristics and debugging techniques of structures

Learning objectives
- Understand the significance of custom data types
- Know steps and methods of type definition, variable definition, initialization, and reference of structures
- Know the concept of unions and how to use them
- Know the concept of enumerations and how to use them

3.1 Concept of structures

3.1.1 Introduction

There were four students in a study group instructed by Mr. Brown. Their information was recorded in a student management table, as shown in Figure 3.1. One day, Mr. Brown asked his students, "We have learned how to compute total grade in a two-dimensional table, can you write a program that computes total grades and prints the entire management table?"

ID	Name	Gender	Admission Year	Computer architecture	C	Compiler	Operating System	Total
1001	ZhaoYi	M	2009	90	83	72	82	
1002	QianEr	M	2009	78	92	88	78	
1003	SunSan	F	2009	89	72	98	66	
1004	LiSi	F	2009	78	95	87	90	

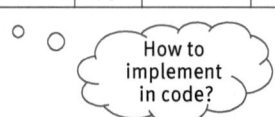

Figure 3.1: Student management table.

https://doi.org/10.1515/9783110692303-003

Compared with a two-dimensional array, data in this table are not of the same type. To process this table, we must first figure out how to access the data on a computer before designing an algorithm. To be more specific, we need a method to store the table in the computer and to retrieve grades from the table. This is also the general approach to solve problems from a computer's perspective.

3.1.2 Storage solution of mixed data table

3.1.2.1 Discussion of possible storage solution of mixed data table

Let us discuss possible solutions to store a mixed data table.

Based on the characteristics of the table and concept of array storage we have learned, we can use two methods to store the table: by row or by column. The pros and cons of these two methods are shown in Figure 3.2.

Solution	Characteristics	Issues
By column	– Each column is a 1-d array – Has existing solution	– Computation involves multiple 1-d arrays – Computation is inconvenient
By row	– Multiple types of data in each row – Easier to access using row offsets – Consistent with our experience	– No existing storage and processing solution – Can use storage solution of arrays

An array is a collection of variables of the same type "By row: combinatorial data structure"is a group of variables of different types

Figure 3.2: Possible solutions to table storage.

Of course, we can store each column in a one-dimensional array. However, total grade computation will then involve multiple one-dimensional arrays. It is tedious to do so in programs. If the computation is done in a module, then there is no easy way to pass information from a row down to child functions.

On the other hand, if we store the table by row, data entries of one individual are stored consecutively, so the sum can be computed in the same way as it is done in one-dimensional arrays. By encapsulating the sum computation in a module, we only need to pass row addresses to it, as we have done in two-dimensional arrays.

To sum up, it is easier to process data if we store the table by row.

3.1.2.2 Issues of constructing "combinatorial data"

There are multiple rows in a table, but they are essentially a repetition of a single row. As a result, it suffices to figure out how we should store a single row. Hence, the key is to "pack" data of different types together and store them in a continuous space, whose beginning address functions as a reference to this space.

We can now list all given conditions and our expectations of the new storage solution as follows:

- It can store multiple data entries, each of which may have its own type.
- Users can determine the number of data entries and values.
- Data mentioned earlier are "packed" together as one entity and stored in a continuous space.
- Each data entry can be accessed independently.

3.1.2.3 Key elements of constructional data

An array is a group of variables of the same type. However, a "combinatorial data structure" requires us to reconstruct new concepts and methods based on arrays. Using these new methods, we can construct "constructional data."

Based on three key elements of data storage, we can analyze combinatorial data from the perspective of storage size, memory allocation, and data access. As shown in Figure 3.3, they are determined by type, definition, and reference of "combinatorial data," respectively.

Three key elements of data storage			
Storage size	Type of "combinatorial data"	Type size Sum of size of each data entry	Type name keyword+identifier
Memory allocation	Variable definition of "combinatorial data"	Memory is allocated by the system based on types of custom data Multiple data entries are stored consecutively	
Data reference	Variable reference of "combinatorial data"	Reference of single data entry, multiple entries as a whole entity and address	

What issues we need to consider when constructing "combinatorial data"?

This identifier is user-defined

Figure 3.3: Key elements of storage of constructional data.

(1) Type of constructional data

The storage size is determined by data type. A type is identified by its name and size.

The system cannot predict data in the mixed data table because they are generated by users. Thus, users need to "construct" a type for the table on their own.

There are multiple types of data in the table, so the size of the table type should be the sum of the sizes of each entry.

Because such a combinatorial type is data-dependent, its size varies for different tables. As a result, it is not possible to use a single type for all of them. Otherwise, the system cannot allocate a suitable amount of memory for each table. Hence, it is the programmers' task to define types. It is thus necessary to design syntax for type definitions. In C language, such definitions are done in the format "keyword + identifier," in which programmers name the identifier.

(2) Definition of constructional data

After defining a variable for "combinatorial data," the system should allocate memory based on the custom type. The data entries should be stored continuously.

(3) Reference of constructional data

To retrieve data, programmers should be able to access a single data entry, all entries as a whole entity or the address of the "combinatorial data" variable.

"Combinatorial data" are called structures in C. A structure is a collection constructed by data of different types. Structures in C make storage and processing of complex data structures possible.

3.2 Storage of structures

3.2.1 Type definitions of structures

Figure 3.4 shows the definition of a structure (struct) and its data entries. Structures are one of the aggregate data types in C.

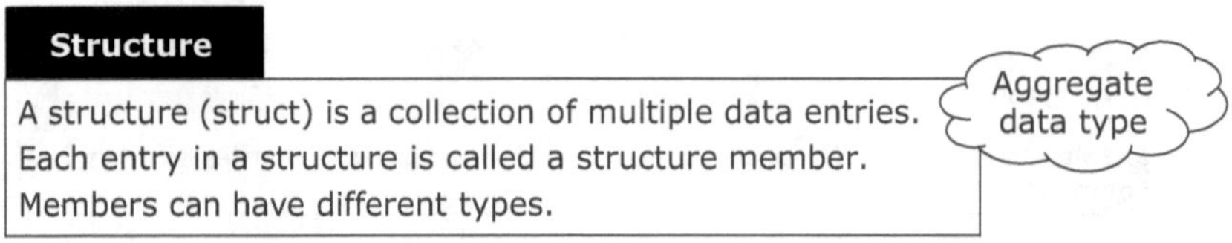

Structure

A structure (struct) is a collection of multiple data entries. Each entry in a structure is called a structure member. Members can have different types.

Aggregate data type

Figure 3.4: Definition of structures.

Figure 3.5 presents some concepts related to the type of structures.

Structure names are identifiers defined by programmers to reference structures conveniently. The type name of a structure consists of keyword struct and the structure name. A structure definition consists of type names and definitions of structure members. Although structure names are optional, it is not recommended to omit them. A structure must be defined before being used. Members of a structure can be of any valid types in C.

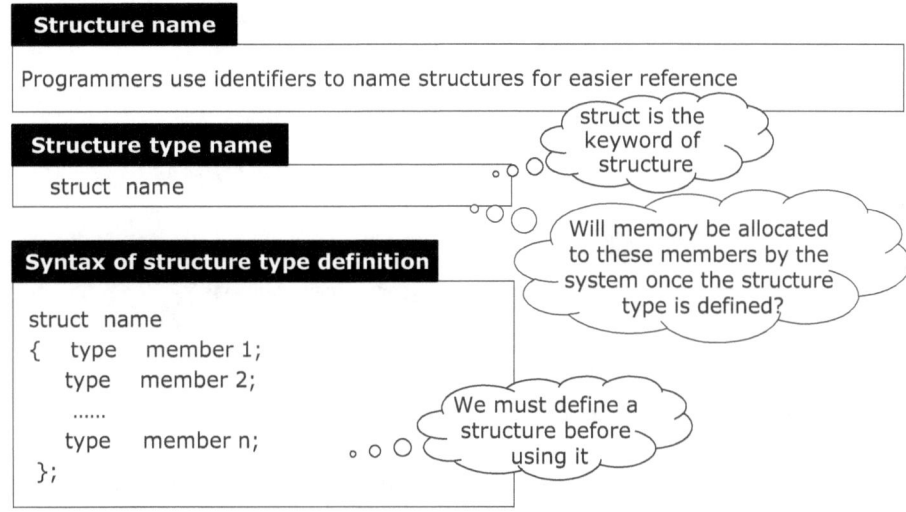

Figure 3.5: Concepts related to structure type.

Think and discuss Will the system allocate memory for structure members after defining the structure?
Discussion: Note that a structure is a user-defined data type. In C, types describe the size of memory allocated, but a type definition will not trigger memory allocation. Memory will not be allocated until a variable of this type is defined.

Example 3.1 Structure definition of student management table
Define a structure for data in Figure 3.6

ID	Name	Gender	Admission year	Computer architecture	C	Compiler	Operating system	Total
id	name	gender	time	score_1	score_2	score_3	score_4	total

Figure 3.6: Student management table.

Analysis
Figure 3.7 presents two solutions to structure definition for the table.

Solution 1: the structure name is student, which becomes the type name together with keyword struct. The members are defined one by one, each with an appropriate type.
Solution 2: we can combine data of the same type into an array to make the definition simpler.

```
                  Solution 1
struct student
{
    int    id;
    char   name[10];                                    Solution 2
    char   gender;          struct student
    int    time;           {
    int    score_1;            int  id;
    int    score_2;            char name[10];
    int    score_3;            char gender;
    int    score_4;            int  time;
    int    total;             int  score[4]; //Combine 4 grades in an array
} ;                            int  total;
                          }
```

Figure 3.7: Solutions of structure definition for the student management table.

3.2.2 Definition of structure variables

With the definition of a structure type, we can define structure variables. As shown in Figure 3.8, the definition is similar to plain variable definitions, except that the type is a structure type.

Syntax of structure variable definition

structureType   variableName;

Figure 3.8: Definition of structure variable.

Let us consider the following example:

Example 3.2 Variable definitions related to student management table
There were 30 students in Mr. Brown's class, whose information is recorded in the student management table format shown earlier. Please write out definitions of the following variables:
– a structure variable;
– an array of 30 structure variables;
– a pointer pointing to a structure object.

Analysis
Figure 3.9 shows required definitions, where struct student is the structure type, x is the name of the structure variable, com[30] is the structure array, and sPtr is the pointer pointing to a structure.

ID	Name	Gender	Admission year	Computer architecture	C	Compiler	Operating system	Total
id	name	gender	time	score_1	score_2	score_3	score_4	total

Description	Form
Structure type	struct student
Structure variable definition	struct student x;
Structure array definition	struct student com[30];
Structure pointer definition	struct student *sPtr;

Figure 3.9: Variable definitions related to the student management table.

3.2.3 Structure initialization

Similar to arrays, a structure variable can also be initialized, as shown in Figure 3.10.

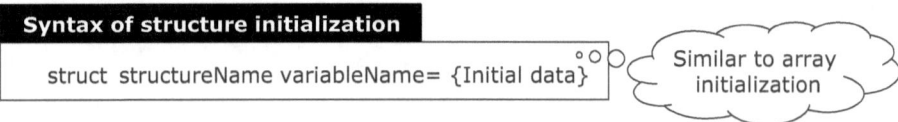

Syntax of structure initialization

struct structureName variableName= {Initial data}

Similar to array initialization

Figure 3.10: Syntax of structure variable initialization.

Example 3.3 Initialization of structure array
Please initialize structure array com[30] with data in the student management table.
Analysis
As shown in Figure 3.11, we shall only initialize the first four rows as a demonstration. The unin-
itialized elements will be set to 0 by the system automatically.
  Now we can store data of various types in a single data structure.

ID	Name	Gender	Admission year	Computer architecture	C	Compiler	Operating system	Total
1001	ZhaoYi	M	2009	90	83	72	82	
1002	QianEr	M	2009	78	92	88	78	
1003	SunSan	F	2009	89	72	98	66	
1004	LiSi	F	2009	78	95	87	90	

```
//Structure array initialization
struct   student com [30]
={ { 1001, "ZhaoYi", 'M', 2009, 90, 83, 72, 82 },
    { 1002, "QianEr", 'M', 2009, 78, 92, 88, 78 },
    { 1003, "SunSan", 'F', 2009, 89, 72, 98, 66 },
    { 1004, "LiSi", 'F', 2009, 78, 95, 87, 90 }
  };
```

Uninitialized elements are set to 0 by the system

Figure 3.11: Initialization of structure array.

3.2.4 Memory allocation of structure variables

Memory is allocated to variables based on their definitions. We shall examine how memory is allocated to structure variables using objects we defined previously, including structure variable x, structure pointer sPtr, and structure array com.

3.2.4.1 Definitions related to structure

```
struct student     //Structure type definition
{
    int id;
    char name[10];
    char gender;
    int time;
    int score[4];
    int total;
};
struct student x;   //Structure variable definition
struct student com [10]   //Structure variable definition and initialization
    ={{1001, "ZhaoYi", 'M', 2009, 90, 83, 72, 82 },
        {1002, "QianEr", 'M', 2009, 78, 92, 88, 78 },
        {1003, "SunSan", 'F', 2009, 89, 72, 98, 66 },
        {1004, "LiSi", 'F', 2009, 78, 95, 87, 90 }
        };
struct student *sPtr; //Structure pointer definition
sPtr=com; //Make the pointer point to array
x=com[2]; //Assign com[2] to x
```

3.2.4.2 Memory layout of structure variables

Figure 3.12 shows memory layout of structure variables.

Size of memory allocated to structure variable x is the sum of size of memory required by each member in the structure.

Structure array com has 10 rows, each of which has the same size as structure variable x.

Using assignment statement sPtr = com, we point sPtr to the beginning address of array com. This is possible because they are of the same type. To make the pointer point to com[9], we can simply move it by nine elements with statement sPtr + 9.

3.2.4.3 Inspection of memory layout of structure variables

Figure 3.13 shows the memory layout of these variables in debugger windows.

Structure array com has 10 elements, each of which can be expanded by clicking the plus sign in front of it. Here we have only expanded com[0]. It is clear after

struct student x,   com [10],   *sPtr;

x	ID	Name	Gender	Admission year	Grade 1	Grade 2	Grade 3	Grade 4	Total
x									

sPtr=com;

sPtr

	ID	Name	Gender	Admission Year	Grade 1	Grade 2	Grade 3	Grade 4	Total
com[0]									
com[1]									
...	...			...					
com[9]									

sPtr+9

Figure 3.12: Memory layout of variables related to student management table.

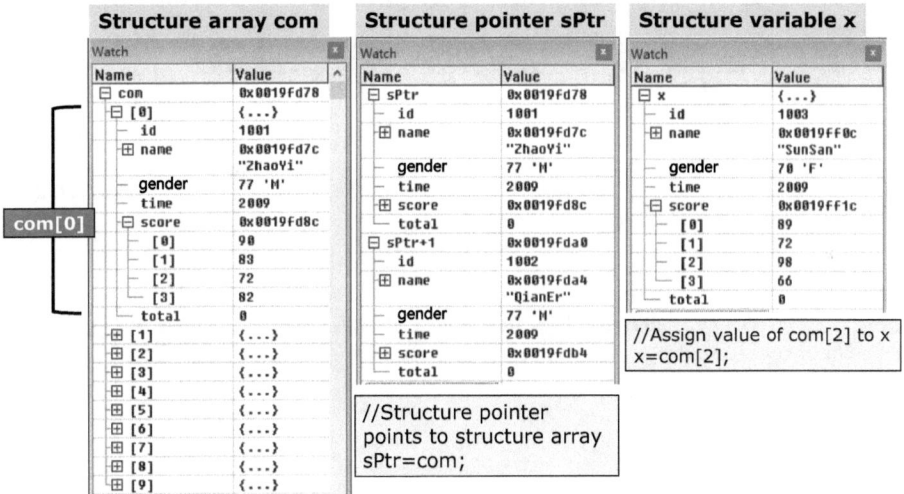

Figure 3.13: Inspection of the memory layout of variables related to the student management table.

expansion that each element consists of data entries defined in the structure-type definition.

The Watch window for variable x shows that assigning value to a structure variable copies all data entries.

In the window for sPtr, we can inspect contents pointed to by it. sPtr points to the beginning address of com, and we can verify that id is indeed 1001. sPtr + 1 has id 1002, so it points to com[1]. This is consistent with the definition of pointer offset.

3.2.4.4 Data alignment of structures

Because members of a structure can have different types, their addresses need to be aligned during memory allocation. Let us look at two examples first.

Example 3.4 Data alignment for basic types
Suppose we have three structure variables A, B, and C, and we know their initial values. Also, suppose the size of short and size of long are 2 and 4 bytes, respectively, in the runtime environment. After testing, we have obtained lengths of these variables, which are 6, 8, and 8 bytes, respectively, as shown in Figure 3.14. Please explain why this is the case after inspecting the memory.

Suppose:
sizeof(short)=2   sizeof(long)=4

```
struct
{ short  a1;
    short  a2;
    short  a3;
} A ={1,2,3};
```

```
struct
{
    long  a1;
    short  a2;
} B ={4,5};
```

```
struct
{
    short a1;
    long  a2;
} C={6,7};
```

sizeof(A)=?   **sizeof(B)=?**   **sizeof(C)=?**

Why is this the case?

Result:
   sizeof(A)=6   sizeof(B)=8   sizeof(C)=8

Figure 3.14: Data alignment.

Analysis
Figure 3.15 shows the addresses of A's structure members. They all have length of 2 bytes and are stored consecutively.

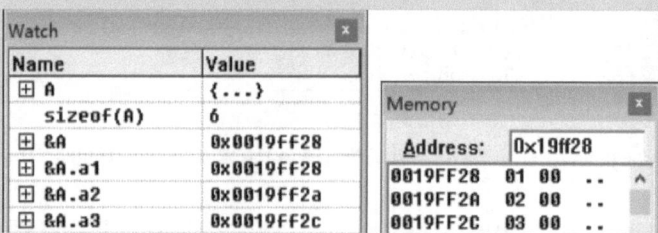

Watch			Memory	
Name	Value		Address:	0x19ff28
⊞ A	{...}			
sizeof(A)	6		0019FF28 01 00 ..	
⊞ &A	0x0019FF28		0019FF2A 02 00 ..	
⊞ &A.a1	0x0019FF28		0019FF2C 03 00 ..	
⊞ &A.a2	0x0019FF2a			
⊞ &A.a3	0x0019FF2c			

Figure 3.15: Memory layout of A.

Figure 3.16 shows the addresses of B's structure members. B.a1 is stored at 0x19ff20 and has 4 bytes. B.a2 is store right after it at address 0x19ff24. sizeof(B) yields 8, so 4 bytes are allocated to B.a2. However, short type only takes up 2 bytes. Thus, the remaining 2 bytes are not in use.

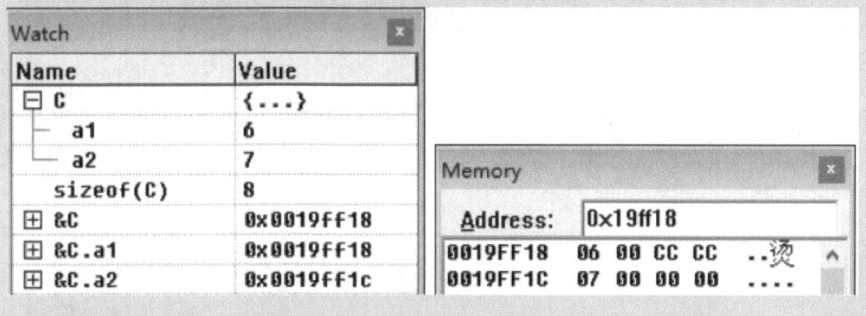

Figure 3.16: Memory layout of B.

Figure 3.17 shows the addresses of C's structure members. C.a1 is stored at 0x19ff18. It is of type short, so its length is supposed to be 2 bytes. Nonetheless, C.a2, a long variable, is stored at 0x19ff1c, which is 4 bytes after C.a1. sizeof(C) yields 8 and we know the length of a long variable is 4 bytes, so 4 bytes are allocated to C.a1. Once again, 2 bytes are not in use.

Figure 3.17: Memory layout of C.

Think and discuss Why are there "holes" in memory that are not used?
Discussion: We can infer from the memory layout shown earlier that these memory units are allocated in such a way that the addresses of structure members are aligned. Data alignment allows the CPU to access memory more efficiently. It is an optimization done by compilers during memory allocation of variables. The optimization (alignment) rule for basic types is as follows:

Variable address %N = 0 (Alignment parameter N = sizeof(variable type))

Note: this rule may vary in different compilers.

Knowledge ABC Memory allocation rules of structures (VC++ 6.0)
1. Member storage order
   Members of a structure are stored in the order in which they are defined. The first member is stored at the lowest address, while the last member is stored at the highest address.
2. Data alignment parameter
   (1) Alignment parameter for a member:
       N = min(sizeof(member type), n)
       Note: the value of n is configurable in VC++ 6.0. Its default value is 8 bytes.
   (2) Alignment parameter for a structure: M = maximum of alignment parameters of all members in the structure

3. Memory allocation rules of structures
   (1) Structure size L: L%M = 0 (empty bytes should be padded if necessary).
   (2) Address of a member x: x%N = 0 (if the size of the member is less than M, the next member is padded).

Memory is allocated in multiples of M bytes: if a member is longer than M bytes, then M more bytes are allocated; if a member is shorter than M bytes, then the next member is padded following the same set of rules (which also apply to nested structures).

Example 3.5 Data alignment for constructional types
Figure 3.18 shows the definition of structure struct stu and its information in the Watch and the Memory windows.
   Let struct stu x = {1, "ZhaoYi", "Male", 3, 4, 5, 6, 7 }.

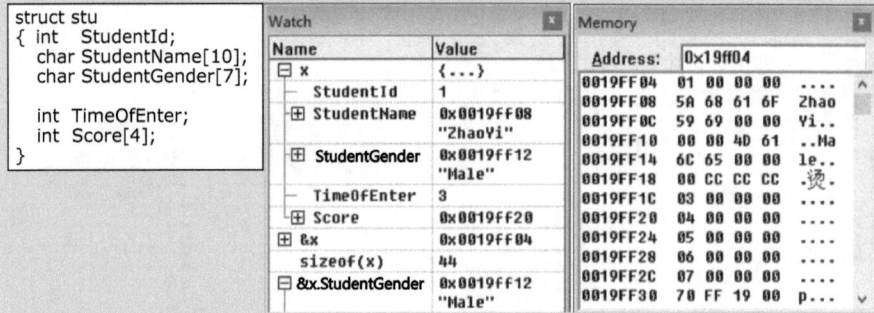

Figure 3.18: Memory layout of variable x.

The length of memory allocated to x is 0x19ff30-0x19ff04=0x2c=44 bytes; sum of length of its members=(int+ char*10+char*7+int+int*4)=41 bytes. The difference between these two values is 3 bytes, so "holes" exist in the memory, as shown in Figure 3.19.

Member	Beginning address	4 bytes			
int StudentId	19FF04	01	00	00	00
char StudentName[10]	19FF08	5A	68	61	6F
		59	69	00	00
char StudentGender[7]	19FF12	00	00	4D	61
		6C	65	00	00
		00	CC	CC	CC
int TimeOfEnter	19FF1C	03	00	00	00
int Score [4]	19FF20	04	00	00	00
		05	00	00	00
		06	00	00	00
		07	00	00	00

Memory "hole" 3 bytes

Figure 3.19: Memory "holes.".

The alignment parameter of structure x is M = sizeof(int) = 4.

Note:

(1) Objects stored in unit 0x19FF10 and 0x19FF11 are StudentName[8] and StudnentName[9], respectively.

(2) The beginning address of StudentGender is 0x19FF12.
     The alignment parameter of StudentGender is N = min(sizeof(member type), 8) = sizeof (char) = 1.
     Because 0x19FF12%N = 0, the 7 elements of StudentGender is stored in units starting from 0x19FF12.

(3) The beginning address of TimeOfEnter is 0x19FF1C.

The alignment parameter of TimeOfEnter is N = sizeof(int) = 4.

The address of the next empty unit after 7 elements of StudentGender is 0x19FF19. None of the numbers in the range 0x19FF19 to 0x19FF1B is multiple of 4, as shown in Figure 3.20, so the beginning address of TimeOfEnter has to be 0x19FF1C, which is a multiple of 4. As a result, aligning TimeOfEnter leads to the 3-byte "hole" after StudentGender.

Watch	
Name	Value
0x12FF69%4	1
0x12FF6A%4	2
0x12FF6B%4	3
0x12FF6C%4	0

Figure 3.20: Result of addresses mod 4.

We can conclude that a good structure member design makes the structure simpler and saves memory space. Carefully designed structures can make our programs more efficient.

3.2.5 Referencing structure members

We obtain memory for structures by defining structure variables and assigning values to them through initialization. These are all write operations of data. Because we need to read them as well, a referencing method is necessary for members of a structure variable.

There are three ways to reference a structure member in C, as shown in Figure 3.21. The first one references a member by its name. Its syntax is structure variable name and member name connected by a dot. The rest references a member by its address. They require a pointer pointing to the structure. This pointer is then used together with member names to complete the task. Essentially, these two methods work in the same way.

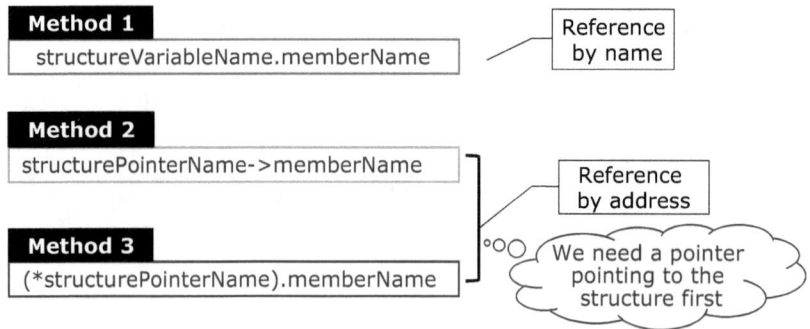

Figure 3.21: Syntax of referencing structure members.

Example 3.6 Example of referencing structure members
Figure 3.22 shows concrete examples of member referencing using structure and variables defined previously.

```
struct student
{
    int  id;
    char name[10];
    char gender;
    int  time;
    int  score[4];
    int  total;
}
struct student
x, com[30],*sPtr;
```

Object		Value to be referenced	Statement	Reference prefix
Structure variable x	Total grade	x.total	x
	The 0th grade	x.score[0]	
Structure array com[30]	Total grade of the 1st student	com[1].total	com[i]
	The 0th grade of the 2nd student	com[2].score[0]	
Structure pointer sPtr	Total grade	sPtr->total	sPtr->
	The 3rd grade	sPtr->score[3]	
	Total grade	(*sPtr).total	(*sPtr)
	The 2nd grade	(*sPtr).score[0]	

Figure 3.22: Example of referencing structure members.

To reference a member of the structure variable x, we use the statement "x.member name."
   To reference a member of structure array com, we use the statement "com[index].member name."
   To reference a member of the structure pointer sPtr, we use either "sPtr-> member name" or "(*sPtr).member name."

3.3 Applications of structures

Example 3.7 Comparison of structures and arrays
Write a program that finds the highest score and corresponding seat number in Figure 3.23, prints the information, and swaps it with the first column.

Seat No.	1	2	3	4	5	6
Grade	90	80	65	95	75	97

Figure 3.23: Data grid.

Analysis

1. Data structure design

We may use one of the following three solutions:

(1) Using a one-dimensional array

Score array: int score [6] = {90,80,65,95,75,97};

Seat number array: int set[6] = {1,2,3,4,5,6};

(2) Using a two-dimensional array

Combination of score and seat number: int score[2][6] = {{90,80,65,95,75,97},{1,2,3,4,5,6}};

We have learnt how to store data with arrays: data of the same type are stored in order; each element is accessed using array name and index.

(3) Using a structure

Solution 1:

```
struct node {
int score[6];
int seat[6];}
struct node x={{90,80,65,95,75,97},{1,2,3,4,5,6}}
```

Solution 2:

```
struct node {
int score;
int seat;}
struct node y[6]={{90,1},{80,2},{65,3},{95,4},{75,5},{97,6}};
```

Structures "pack" correlated data together. Type of a structure is defined by users. Memory is allocated during definition of variables of the structure type.

After storing data into memory using structures, we need to reference them for further computation. Figure 3.24 shows how data are stored and accessed using a one-dimensional array, two-dimensional array, and structure.

Figure 3.25 shows how to reference members of structure variable x and structure array y and their corresponding values.

	Address	Type	Variable reference	Storage order	Characteristics
1-d array	Array name score	int	score[index]	Elements are stored consecutively; two arrays are not necessarily stored consecutively	It is convenient to handle large amount of data of the same type using arrays
	Array name seat	int	seat[index]		
2-d array	Array name score	int	score[index][index]	Elements are stored consecutively in a row-first manner	
Struc-ture	Address of x	struct node	x.score [index] x.seat[index]	Members of the structure are stored consecutively; array score is followed by array seat Members of the structure are stored consecutively; pairs of score and seat are stored in the array consecutively	Structures combine correlated data together so that we can use one single variable to access them
	Address of y	struct node	y[index].score y[index].seat		

Figure 3.24: Data storage and reference.

Storage order of variable x		Storage order of array y[6]	
Member variable	Value	Member variable	Value
x.score[0]	90	y[0].score	90
x.score[1]	80	y[0].seat	1
x.score[2]	65	y[1].score	80
x.score[3]	95	y[1].seat	2
x.score[4]	75	y[2].score	65
x.score[5]	97	y[2].seat	3
x.seat[0]	1	y[3].score	95
x.seat[1]	2	y[3].seat	4
x.seat[2]	3	y[4].score	75
x.seat[3]	4	y[4].seat	5
x.seat[4]	5	y[5].score	97
x.seat[5]	6	y[5].seat	6

Figure 3.25: Storage of x and y.

2. Algorithm design

Figure 3.26 shows pseudo code of the algorithm.

Pseudo code
Find the current maximum in score, record the corresponding seat number as num
Swap max with the 0th element of score
Swap num with the 0th element of seat
Output contents of array score and array seat

Figure 3.26: Algorithm of finding the highest score and swapping it with the first column.

3. Code implementation 1

```
1   //Using 1-d array
2   #include <stdio.h>
3   #define MAX 6
4
5   int main(void)
6   {
7       int score[MAX]={90,80,65,95,75,97};
8       int seat[MAX]={1,2,3,4,5,6};
9       int max, num;
10      int temp1, temp2;
11
12      //Find maximum of score, store it in max and the index in num
13      max=score[0]; //Use the 0-th element as comparison basis
14      num=1;
15      for (int i=1; i< MAX; i++)
16          {
```

```
17    if (max < score[i])
18    {
19       max=score[i];
20       num=seat[i];
21    }
22  }
23
24  //Swap the maximum with the 0-th element
25  temp1=score[0];
26  temp2=seat[0];
27  score[0]=max;
28  seat[0]=num;
29  score[num-1]= temp1;
30  seat[num-1]= temp2;
31
32  //Output
33  printf("No. 1: seat no. %d, %d pts\n", seat[0],score[0]);
34  return 0;
35 }
```

Program result:
No. 1: seat no. 6, 97 pts

4. Code implementation 2

```
1  //Using 2-d array
2  #include <stdio.h>
3  #define MAX 6
4  int main(void)
5  {
6    int score[2][MAX]=
7    { {90,80,65,95,75,97},
8      { 1, 2, 3, 4, 5, 6}
9    };
10   int max, num;
11   int temp1, temp2;
12
13   //Find maximum of score, store it in max and the index in num
14   max=score[0][0]; //Use the 0-th element as comparison basis
15   num=1;
16   for (int i=1; i< MAX; i++)
17   {
18     if (max < score[0][i])
19     {
20        max=score[0][i];
21        num=score[1][i];
22     }
23   }
24
```

```
25    // Swap the maximum with the 0-th element
26    temp1=score[0][0];
27    temp2=score[1][0];
28    score[0][0]=max;
29    score[1][0]=num;
30    score[0][num-1]= temp1;
31    score[1][num-1]= temp2;
32
33    //Output
34      printf("No. 1: seat no. %d, %d pts\n ",score[1][0],score[0][0]);
35      return 0;
36  }
```

Program result:

No. 1: seat no. 6, 97 pts

5. Code implementation 3

```
1   //Using the first structure definition
2   #include <stdio.h>
3   #define MAX 6
4   int main(void)
5   {
6     struct node
7     {
8       int  score[MAX];
9       int  seat[MAX];
10    } x = { {90,80,65,95,75,97}, {1,2,3,4,5,6} };
11      int  max,num;
12      int  temp1,temp2;
13
14      // Find maximum of score, store it in max and the index in num
15      max=x.score[0]; // Use the 0-th element as comparison basis
16      num=1;
17      for (int i=1; i< MAX; i++)
18      {
19      if (max < x.score[i])
20        {
21        max=x.score[i];
22        num=x.seat[i];
23        }
24      }
25
26      // Swap the maximum with the 0-th element
27      temp1=x.score[0];
28      temp2=x.seat[0];
29      x.score[0]=max;
30      x.seat[0]=num;
31      x.score[num-1]= temp1;
```

```
32    x.seat[num-1]= temp2;
33
34    //Output
35    printf("No. 1: seat no. %d, %d pts \n", x.seat[0],x.score[0]);
36    return 0;
37  }
```

Program result:
No. 1: seat no. 6, 97 pts

6. Code implementation 4

```
1   //Using the second structure definition
2   #include <stdio.h>
3   #define MAX 6
4   int main(void)
5   {
6     struct node
7     {
8     int  score;
9     int  seat;
10    } y[6]={{90,1},{80,2},{65,3},{95,4},{75,5},{97,6}};
11    int  max,num;
12    int  temp1,temp2;
13
14    // Find maximum of score, store it in max and the index in num
15    max=y[0].score; // Use the 0-th element as comparison basis
16    num=1;
17    for (int i=1; i< MAX; i++)
18    {
19      if (max < y[i].score)
20      {
21      max=y[i].score;
22      num=y[i].seat;
23      }
24    }
25
26  // Swap the maximum with the 0-th element
27    temp1=y[0].score;
28    temp2=y[0].seat;
29    y[0].score=max;
30    y[0].seat=num;
31    y[num-1].score= temp1;
32    y[num-1].seat= temp2;
33
34  //Output
35    printf("No. 1: seat no. %d, %d pts\n",y[0].seat,y[0].score);
36    return 0;
37  }
```

Program result:
No. 1: seat no. 6, 97 pts

Example 3.8 Printing student management table
In the introduction section of this chapter, Mr. Brown asked his students to print the table on screen.

Analysis
1. Data structure design
We have studied how to store the table in Section 3.2. To be more specific, we shall use the following structure definition for the student management table:

```
struct student
{
  int id;
  char name[10];
  char gender;
  int time;
  int score[4];
  int total;
};
```

To compute the sum, we need to retrieve the score data by referencing structure members.

(1)   Reference by name
Suppose there are i students and j classes, as shown in Figure 3.27. Then a student's grade in one class is com[i].score[j], in which i is the index of structure array com and j is the index of structure member score. i and j control which row and which column we will be accessing.

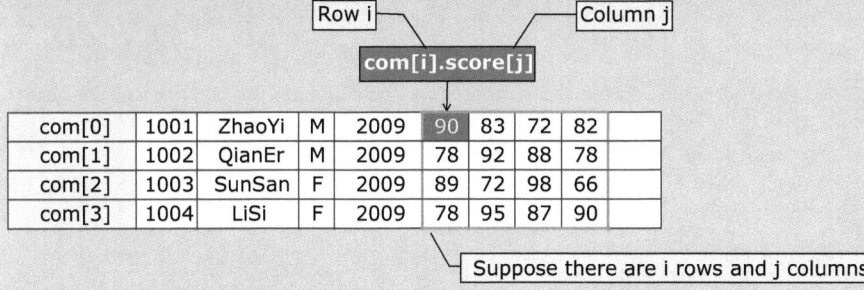

Figure 3.27: Grade reference in student management table 1.

(2)   Reference by address
As shown in Figure 3.28, we point sPtr to the beginning address of com first. Note that the offset of sPtr is one row. To make computation easier, we introduce another pointer ptr that points to a single grade with ptr = sPtr->score. Note that the score here is an array name, so it is also an address. Now we can use sPtr and ptr to control row and column we access.

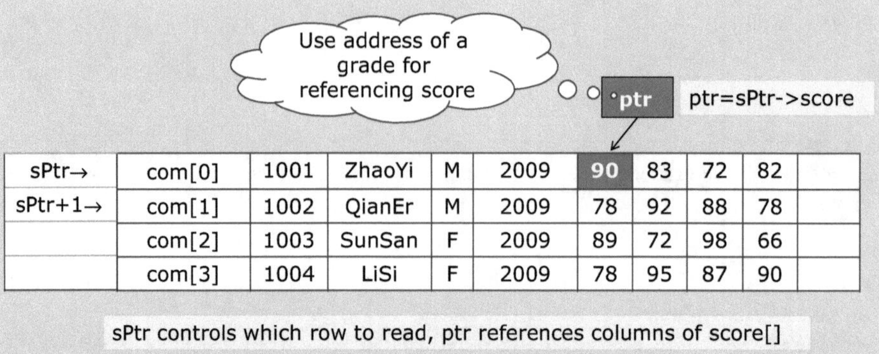

sPtr controls which row to read, ptr references columns of score[]

Figure 3.28: Grade reference in student management table 2.

2. Code implementation
Figure 3.29 shows the reference by name program.

```
01 #include <stdio.h>
02 #define N 4//Number of students
03 #define M 4//Number of courses
04  struct  student
05 {
06      int  id;
07      char name[10];
08      char gender;
09      int  time;
10      int  score[M];
11      int  total;
12 };
13  int main(void)
14 {
15      struct student com [N]
16      = {{ 1001, "ZhaoYi", 'M', 2009, 90, 83, 72, 82 },
17         { 1002, "QianEr", 'M', 2009, 78, 92, 88, 78 },
18         { 1003, "SunSan", 'F', 2009, 89, 72, 98, 66 },
19         { 1004, "LiSi", 'F', 2009, 78, 95, 87, 90 }
20      }; //Structure array initialization
21
22      int i, j;
23      printf( "ID Name Gender Admission Year CompArch C Compil OS Total\n");   //Table header
24      for(i=0; i<N; i++)
25      {
26          com[i].total = 0;
27          for ( j= 0; j< M; j++)
28          {
29              com[i].total +=com[i].score[j];
30          }
31          printf( "%d  %s  %s   %d",com[i].id,com[i].name,&com[i].gender,com[i].time);
32          printf( "   %d   %d",com[i].score[0],com[i].score[1]);
33          printf( "   %d   %d   %d\n",com[i].score[2],com[i].score[3],com[i].total);
34      }
35      return 0;
36 }
```

Structure type definition

Structure array definition and initialization

Compute total grade of a row

Figure 3.29: Student management table processing program 1.

Note:

Lines 4–12 define the structure type.

Lines 15–20 define and initialize the structure array.

Line 23 prints the header of the table.

Lines 27–30 compute the total score for one student.

The for loop on line 24 repeats the sum computation N times.

Lines 31–33 print other data in the row.

Program result:

ID	Name	Gender	AdmissionYear	CompArch	C	Compil	OS	Total
1001	ZhaoYi	M	2009	90	83	72	82	327
1002	QianEr	M	2009	78	92	88	78	336
1003	SunSan	F	2009	89	72	98	66	325
1004	LiSi	F	2009	78	95	87	90	350

The implementation shown in Figure 3.30 uses the same algorithm but references grades by address. Please refer to the program earlier for the first 20 lines.

```
21
22   int i, j;
23   printf(" ID Name Gender AdmissionYear CompArch C Compil OS Total\n ");  //Table header
24   for (i=0; i<N; i++)
25   {
26      com[i].total = 0;
27      for ( j= 0; j< M; j++)
28      {                                    ─┤Compute total grade of a row│
29         com[i].total +=com[i].score[j];
30      }
31      printf( "%d  %s  %s   %d",com[i].id,com[i].name,&com[i].gender,com[i].time);
32      printf( " %d %d",com[i].score[0],com[i].score[1]);
33      printf( " %d %d %d\n",com[i].score[2],com[i].score[3],com[i].total);
34   }
35   return 0;
36 }
```

Figure 3.30: Student management table processing program 2.

Example 3.9 Vote counting machine

There are three candidates in an election, as shown in Figure 3.31. Please write a program to count votes for each candidate. Use keyboard input to simulate the counting process. Each voter has to choose one from the three candidates. Suppose there are N votes in total.

Candidate name	Number of votes
Zhang	
Tong	
Wang	

Figure 3.31: Vote statistics.

Analysis

1. Data structure design

There are two types of data in vote statistics, so it is better to use a structure to store them. The structure should have two members: candidate name and number of votes. There are three candidates, so we can use a structure array to store their information.

(1) Structure design

Information of each candidate can be stored in the following structure:

```
struct person
{ char name[16];   //Candidate name
int sum;           //Number of votes
}
```

(2) Vote statistics table design

There are three candidates, so we use an array of size 3. Each element is initialized with the candidate name and 0 votes.

```
struct person vote[3] ={"Zhang",0, "Tong",0, "Wang",0};
```

2. Algorithm design and code implementation (see Figure 3.32)

Pseudo code
while counter<number of total votes N
Input candidate name in_name
Look for in_name in the statistics table, Add 1 to corresponding number of votes if the person exists in table
Output result

Figure 3.32: Algorithm.

```
//Vote counting#include <stdio.h>
#include <string.h>
#define N 50           //Number of votes
struct person
{ char name[20];       //Candidate name
int sum;               //Total votes
};
int main(void)
{
  struct person vote[3]
   ={"Zhang",0, "Tong",0, "Wang",0};
  int i,j;
  char in_name[20];
  for(i=0;i<N;i++)                //N votes
```

```
{
scanf("%s",in_name);       //Input candidate name
for(j=0;j<3;j++)           //Add one to corresponding sum
if (strcmp(in_name, vote[j].name)==0)
{
vote[j].sum++;
}
}
for (i=0;i<3;i++)          //Output result
{
printf("%s,%d\n",vote[i].name,vote[i].sum);
}
return 0;
}
```

Program reading exercise Finding the eldest person

The following program uses a structure to store names and ages of multiple individuals, to find the eldest person, and to output the result:

```
#define N 4
#include "stdio.h"
static struct man
{
    char name[8];
    int age;
} person[N]= {"li",18,"wang",19,"zhang",20,"sun",22};
int main(void)
{
    struct man *q,*p;
    int i,m=0;
    p=person;
    for (i=0; i<N; i++)
    {
      if (m < p->age) q=p++;
          m=q->age;
    }
    printf("%s,%d",(*q).name,(*q).age);
    return 0;
}
```

3.4 Union

3.4.1 Introduction

There is a lab in the university Mr. Brown works for. The lab is available to members of all related research groups. One should book the lab before using it. However, researchers from different groups cannot use the lab together. We can list research groups and researchers that are entitled to use the lab as follows:

```
Public lab
{
      Research group 1:   Person 1;
      Research group 2:   Person 2;
      . . .
      Research group n:   Person n;
}
```

To save memory space, we also use such a memory sharing strategy in computers. We can store variables that cannot be accessed simultaneously into one memory unit. Such a data structure is called a "union." When we have multiple variables and use exactly one of them each time, we can use a union to store them into the same memory unit.

3.4.2 Memory layout of unions

Similar to structures, type definition, variable definition, and member access are also key issues for unions.

3.4.2.1 Union-type definition

Figure 3.33 illustrates the syntax of the union-type definition.

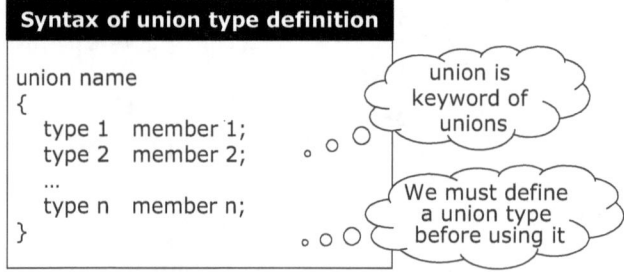

Figure 3.33: Syntax of union-type definition.

Similar to a struct-type definition, a union-type definition merely declares the type. No memory space is allocated at this stage.

3.4.2.2 Union variable definition

Figure 3.34 shows the syntax of defining a union variable, which is similar to that of a structure variable.

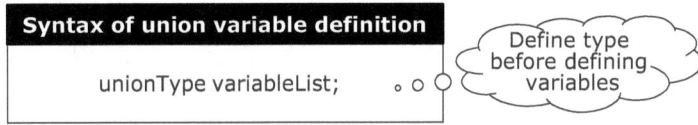

Figure 3.34: Syntax of defining a union variable.

3.4.2.3 Union member reference

Figure 3.35 shows how to reference a union member. We have seen similar syntax in structures.

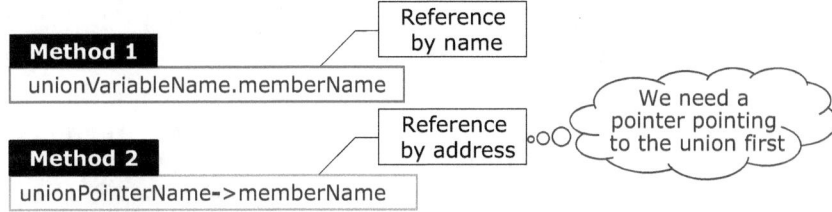

Figure 3.35: Syntax of referencing a union member.

Example 3.10 Memory Layout of Union Members
Suppose we have a union defined as shown in Figure 3.36. Unlike struct members, x, ch, and y have the same address. The length of memory space allocated is determined by the member with the largest size.

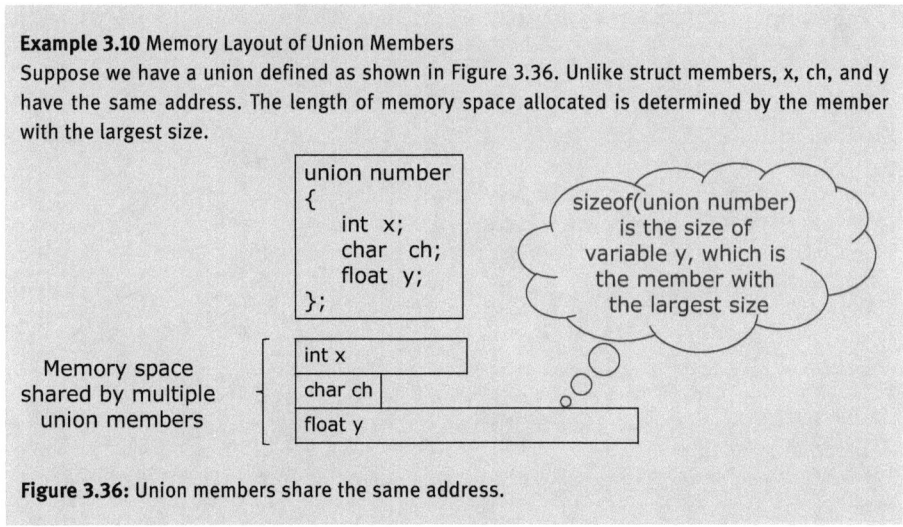

Figure 3.36: Union members share the same address.

3.4.2.4 Comparison of unions and structures

Unions and structures have multiple features in common. Figure 3.37 presents a comparison of these two data structures.

	Union	Structure
Memory size	Memory space is shared by all members, its size is determined by the member with the largest size	Memory space size is sum of sizes of members
Member relation	Only one member is valid at a given time, which is the last stored member	All members are stored consecutively in the order in which they are defined
Relation	Union types can appear in structure type definitions	

Figure 3.37: Comparison of unions and structures.

Example 3.11 Simple program using union
Please examine the memory layout of a union in the debugger.

1. Test program
The following test program defines a union and assigns values to its members in the order in which they are defined. After running the program, we can inspect the memory layout of the union using a debugger:

```c
#include <stdio.h>
int main(void)
{
  union number     //Define a union type
  {
    int x;
    char  ch;
    float y;
  };
  union number unit;    //Define a union variable
  unit.x=1;       //Reference union members
  unit.ch='a';
  unit.y=2;
  return 0;
}
```

2. Debugging
Figure 3.38 shows that the three members are all stored at 0x12ff7c. In particular, it shows memory layout after value 1 is assigned to x, whose value is shown in the Memory window as well.

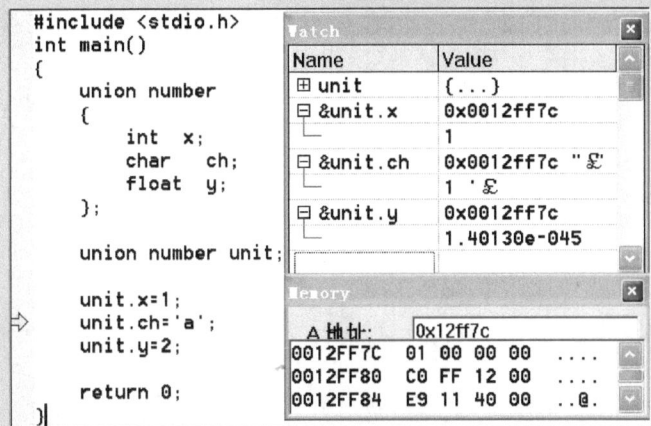

Figure 3.38: Inspection of the memory layout of a union step 1.

Figure 3.39 shows the memory layout after the character 'a' is assigned to ch. ch's value in the Memory window is 0x61, which is exactly the ASCII value of 'a'. This indicates that the valid value of the union has changed to 0x61.

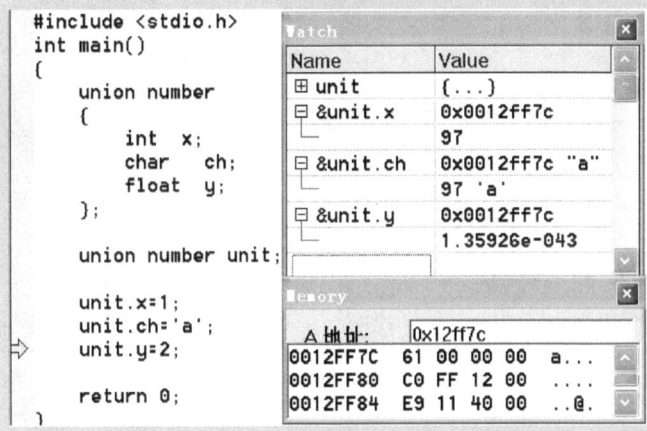

Figure 3.39: Inspection of the memory layout of a union step 2.

In Figure 3.40, we have assigned real number 2 to y. However, the value displayed in the Memory window is 0x40000000. Why is this the case?

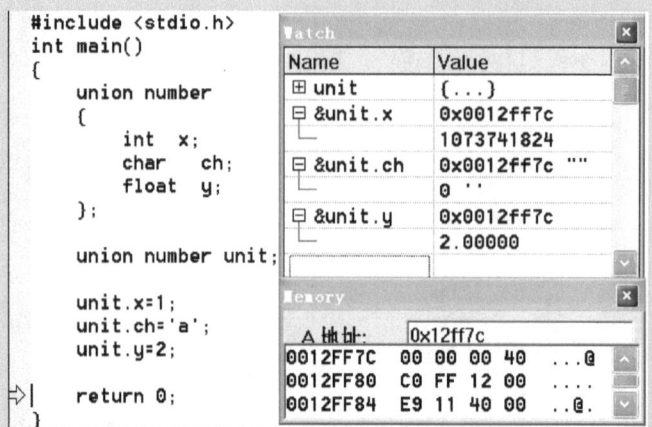

```
#include <stdio.h>
int main()
{
    union number
    {
        int   x;
        char  ch;
        float y;
    };

    union number unit;

    unit.x=1;
    unit.ch='a';
    unit.y=2;

    return 0;
}
```

Watch		
Name	Value	
⊞ unit	{...}	
⊟ &unit.x	0x0012ff7c	
└	1073741824	
⊟ &unit.ch	0x0012ff7c	""
└	0	' '
⊟ &unit.y	0x0012ff7c	
└	2.00000	

Memory	
A 址址:	0x12ff7c
0012FF7C	00 00 00 40 ...@
0012FF80	C0 FF 12 00 ....
0012FF84	E9 11 40 00 ..@.

Figure 3.40: Inspection of the memory layout of a union step 3.

Think and discuss Display format of floating-point variables
The value of the float variable is 2, but why is it displayed as 0x40000000?
Discussion: According to the IEEE754 standard, which we have introduced in the chapter "Basic Data," real number 2 is exactly 0x40000000 if stored as a 32-bit float type number, as shown in Figure 3.41.

Decimal	Normalization	Exponent	Sign	8 bits biased exponent(exponent+127)	23 bits fraction
2	1.0×2^1	1	0	100 0000 0	000 0000 0000 0000 0000 0000

Figure 3.41: Storage format of real number 2.

Program reading exercise Operations on unions
Suppose we have data from multiple teachers, as shown in Figure 3.42. The data include their ID, name, title, number of courses they are teaching (if the title is Lecturer), or number of papers they have published (if the title is Professor). Please write a program that prints these data and computes the total number of papers published.

No.	Name	Title	Number of courses or papers
1	Zhao	L	program
2	Qian	P	3
3	Sun	P	5
4	Li	L	English
5	Zhou	P	4

Figure 3.42: Data for union.

```
1 //Operations on union
2 #include <stdio.h>
3 #define N 5 //Number of teachers
4
5 union work
6 { char course[10]; //Course name
7  int  num; //
8 };
9
10 struct  teachers
11 { int  number; //ID
12  char name[8]; //Name
13  char position; //Title
14  union work x; //Number of courses or papers
15 } teach[N];
16
17 int main(void)
18 {
19   struct teachers teach[N]
20   ={ {1, "Zhao",'L',"program"},
21   {2, "Qian",'P',3},
22   {3, "Sun",'P',5},
23   {4, "Li",'L',"English"},
24   {5, "Zhou",'P',4},
25   };
26   int sum=0;
27
28   for(int i=0; i<N; i++)
29   {
30   printf ( " %3d %5s %c ",teach[i].number,
31      teach[i].name,  teach[i].position);
32   if (teach[i].position =='L')
33   {
34    printf ("%s\n", teach[i].x.course);
35   }
36   else if ( teach[i].position =='P' )
37   {
38  printf ("%d\n", teach[i].x.num);
39    sum=sum+teach[i].x.num;
40   }
41   }
42   printf ("paper total is %d\n", sum);
43   return 0;
44 }
```

Program result:
1 Zhao L program
2 Qian P 3

```
3 Sun P 5
4 Li L English
5 Zhou P 4
paper total is 12
```

3.5 Enumeration

3.5.1 Introduction

When Daniel started to learn watercolor painting, he was shocked by how different colors could be mixed into a new color. He kept asking Mr. Brown questions like "What is the result of mixing red and blue?" or "What if I mix yellow and red?" which made Mr. Brown exhausted. As a result, Mr. Brown decided to write a program that could answer these questions, for given input from his son.

Figure 3.43 shows results of mixing two of three primary colors.

	Red	Yellow	Blue
Red	Red	Orange	Purple
Yellow	Orange	Yellow	Green
Blue	Purple	Green	Blue

number	0	1	2	3	4	5
color	Red	Yellow	Blue	Orange	Purple	Green
string	red	yellow	blue	orange	purple	green

Figure 3.43: Color mixer.

Before he could write the code, Mr. Brown needed to design a data structure. He used a pointer array to store color names in Figure 3.43: char *ColorName[] = {"red","yellow","blue","orange","purple","green"};

The two-dimensional array in Figure 3.43 should be initialized with indices of colors in array ColorName, instead of actual name strings, because indices require less memory space and are easier to process.

```
int ColorTab[3][3]={{0,3,4},{3,1,5},{4,5,2}};
```

During the initialization process, Mr. Brown found that it was difficult to remember the number corresponding to a color. If there were more colors, it would be even harder to remember them and initialize the array correctly. The reason is that names of colors are more intuitive compared with abstract numbers. To solve this issue, Mr. Brown tried to define macros for each color so that he could directly use these intuitive names in the program.

```
int int ColorTab[3][3]={{red,orange,purple},{orange,yellow,green},{purple, green,
blue}};
```

The complete program is as follows:

```
01 #include "string.h"
02 #include "stdio.h"
03 #define red 0
04 #define yellow 1
05 #define blue 2
06 #define orange 3
07 #define purple 4
08 #define green 5
09
10 //Define the color mixer
11   int ColorTab[3][3]={{red,orange,purple},{orange,yellow,green},{purple, green,
blue}};
12
13 int main(void)
14{
15   char color1[8]; //Read input 1
16   char color2[8]; //Read input 2
17   char *ColorName[]= {"red","yellow","blue","orange","purple","green"};
18   int i=0,j=0;
19
20   printf("Please enter any two colors of red, yellow and blue:\n");
21   gets(color1);
22   gets(color2);
23   while (0!=strcmp(color1,ColorName[i])) i++;
   //Find index i of the first input
24   while (0!=strcmp(color2,ColorName[j])) j++;
   //Find index j of the second input
25   //Find mixing result using i and j in the color mixer
26   printf("%s+%s=%s\n",ColorName[i],ColorName[j],ColorName[ColorTab[i][j]]);
27
28   return 0;
29 }
```

However, Mr. Brown needed to define 6 macros for mixing results of three primary colors. If there were more base colors, it would be tedious to define a macro for each possible outcome.

We often use numbers to represent states in programs, but numbers are less intuitive and readable than state names, as we have just seen in the color example.

If we could find a way to represent states using meaningful words in programs, it would be easier to read and understand them.

3.5.2 Concept and syntax of enumeration

In fact, C and some other languages do provide a method of using words in natural languages to represent possible values of a variable. This method is enumeration (enum).

In C, an enumeration is a collection of integer constants represented by identifiers. The value of an enumeration variable must be a member of this collection. It is worth noting that the system will not throw an error if an enumeration variable is assigned a value that is out of the enumeration range.

The syntax of enumerations is similar to that of structures and unions, as shown in Figure 3.44. We can define an enumeration for days in a week as follows:

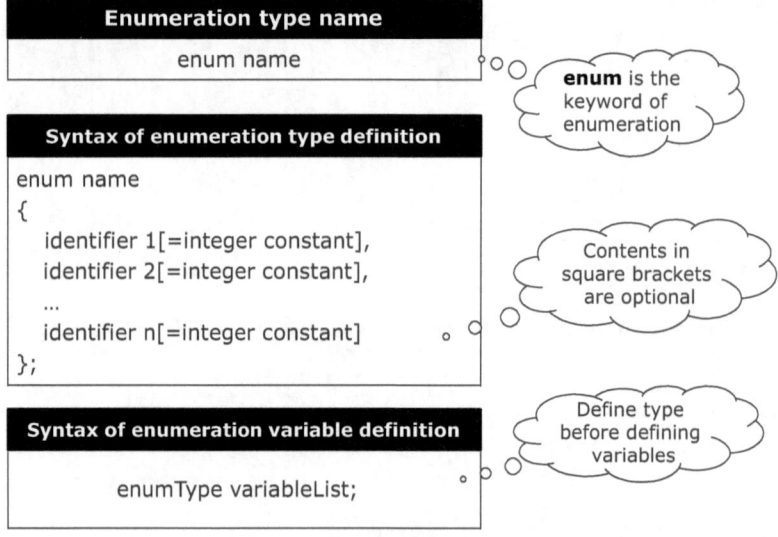

Figure 3.44: Syntax of defining enumeration type and enumeration variable.

```
enum WeeksType {Mon, Tues, Wed, Thurs, Fri, Sat, Sun} ;
enum WeeksType Day;
```

WeeksType is the enumeration type name, Day is an enumeration variable, identifiers in curly brackets are all possible enumeration constants.

Notes:

(1) Identifiers in an enumeration-type definition are constants.

(2) One needs to list all members when defining an enumeration.

(3) Contents in square brackets are optional. If we omit them, numbers 0, 1, 2, . . . will be assigned to the identifiers. However, if one of the members is explicitly assigned a value, members after it will automatically obtain a value, in which each member is one larger than the previous.

We can explicitly assign values to all enumeration members. Note that the values must be integers. For example: enum WeeksType {Mon = 1, Tues = 2, Wed = 3, Thurs = 4, Fri = 5, Sat = 6, Sun = 7};

Besides, we can also explicitly assign values to a few members: enum WeeksType {Mon = 1, Tues, Wed = 1, Thurs, Fri, Sat, Sun}; In this definition, Mon and Wed are defined to be 1. Based on the note earlier, values of Tues, Thurs, Fri, Sat and Sun are 2, 2, 3, 4 and 5, respectively.

(4) Value of an enumeration variable must be one of the enumeration members. For example, it is valid to write statement Day = Wed.

3.5.3 Example of enumerations

Example 3.12 Color mixer using enumeration
Mr. Brown revised his program of the color mixer using an enumeration of colors:

```
#include "string.h"
#include "stdio.h"
//Define enumeration for three primary colors and mixed colors
enum Color{red,yellow,blue,orange,purple,green};
//Define color mixer
int ColorTab[3][3]={{red,orange,purple},{orange,yellow,green},{purple,green,blue}};
int main(void)
{
//Same as before
}
```

Enumerations are similar to macros. Macros replace identifiers with corresponding values in the preprocessing phase, while enumerations do the replacement during compilation. We can consider enumerations as macros in the compilation phase. More on macros can be found in the chapter "Preprocessing."

Example 3.13 Price management
A supermarket often launches discount campaigns. It may offer a different discount for a product during different periods. Please write a program to implement this model.

Analysis
We can list all periods in an enumeration:
```
enum enumType{Time1, Time2, Time3} rebateTime ;
```

Then we can handle different cases using a switch statement:
```
scanf("%d", &rebateTime);
switch (rebateTime)
{
   case Time1:{. . .;break;}
   case Time2:{. . .;break;}
   case Time3:{. . .;break;}
   default:break;
}
```

The code implementation is as follows:
```
#include<stdio.h>
int main(void)
{
  enum enumType{Time1=3, Time2=5, Time3=6};
  float x=1.0;
  int weekday;
  scanf("%d", &weekday);
  switch (weekday)
  {
    case Time1: x=0.5; break;
    case Time2: x=0.8; break;
    case Time3: x=0.9; break;
     default: break;
  }
  printf("Day %d, discount is %f",weekday,x);
  return 0;
}
```

3.5.4 Rules of enumerations

There are many restrictions on enumerations for their uniqueness. We shall use the following enumeration in the discussion:

```
enum WeeksType {Monday, Tuesday, Wednesday, Thursday, Friday, Saturday, Sunday};
enum WeeksType Weekday
```

3.5.4.1 We cannot assign values of other types to an enumeration variable

For example, it is not valid to write Weekday = 10;

Note: this is because 10 is not a member of the enumeration. However, we can assign values of other types to an enumeration variable through forced-type conversion.

3.5.4.2 Arithmetic operations are not allowed on enumeration variables

For example:

```
Weekday = Sat;
Weekday++; //Invalid
```

Note: this is because increment may break the first rule. In this example, Weekday is assigned the last value of enumeration members, so increment will make this value invalid.

3.6 Type definitions

3.6.1 Introduction

3.6.1.1 Porting of music files

Daniel received a music player as a birthday gift, and he spent lots of time listening to music with it. One day, he asked his father, "These songs were stored in the computer, how do they 'fit' into this little box then?" "Ha-ha, that's a great question," answered Mr. Brown.

We can play music on various devices now, but how are music files stored in them? We know data are stored in computers as binary data, so are music files. WAV (Waveform Audio File Format) is one of the most frequently used multimedia audio file formats on PC. It is a digital audio format used to store audio waves, designed by Microsoft and IBM. It was first introduced in Windows 3.1 in 1991. After multiple revisions, it can be used in many operating systems, including Windows, Macintosh, and Linux.

A WAV file has a file header which contains meta information of the file followed by actual music data. Figure 3.45 shows meta information in the header, which consists of multiple data entries. The size of each entry is fixed and does not change across platforms.

We have mentioned before that the sizes of basic types are platform dependent. As a result, programmers need to be careful when porting code to other platforms. When using code processing WAV files on different platforms that use different sizes for int type, we have to modify every integer definition to make sure lengths of data entries in the header are consistent with the standard. This makes code porting difficult.

	Offset	Byte	Type	Content
	00H	4	Char	"RIFF" sign
	04H	4	int32	File size
	08H	4	Char	"WAVE" sign
	0CH	4	Char	"fmt" sign
	10H	4	Char	Transition bytes
	14H	2	int16	Audio format
Header	16H	2	int16	Number of channels
	18H	2	int16	Sample rate
	1CH	4	int32	Byte rate
	20H	2	int16	Block alignment
	22H	2	Char	Bits per sample
	24H	4	Char	"data" sign
	28H	4	int32	Sound data size

Figure 3.45: Format of WAV file header.

In this case, if we rename 32-bit and 16-bit int types as UIN32 and UIN16, we can define the WAV file header as the following structure. When porting the code, it suffices to replace UIN32 and UIN16 with the corresponding types of the target platform:

```
struct tagWaveFormat
{
    char cRiffFlag[4];
    UIN32 nFileLen;
    char cWaveFlag[4];
    char cFmtFlag[4];
    char cTransition[4];
    UIN16 nFormatTag ;
    UIN16 nChannels;
    UIN16 nSamplesPerSec;
    UIN32 nAvgBytesperSec;
    UIN16 nBlockAlign;
    UIN16 nBitNumPerSample;
    char cDataFlag[4];
    UIN16 nAudioLength;
};
```

The replacement can be done using macros.

3.6.1.2 Cases where macros are not enough

Let us examine a special case. We intended to define a and b as integer pointers with the first two lines of the following code. However, it turns out that b is not a pointer, as shown in line 3. This is due to the rule of macro replacement:

```
01 #define PTR int*
02 PTR a, b;
03 int *a, b;
```

Syntactically, line 2 is similar to a variable definition. One may imagine that this issue can be solved if PTR is a data type equivalent to int*. As a result, we need a way to define aliases for data types in C.

3.6.1.3 Define aliases for types

The keyword of the alias definition in C is typedef (type + define). We can rewrite the first two lines mentioned earlier using typedef:

```
01 typedef (int *) PTR
02 PTR a, b;
03 int *a, *b;
```

The purpose of using typedef is to fix issues made by macros and to make code more readable. In practice, typedef is often found in network code and drivers where type sizes are critical. To conform to different compilers, we better define and use our own types. Thus, it suffices to update a few header files when porting our code to new platforms. Typedef can hide complicated structures or platform-dependent data types so that programs are easier to port and maintain.

The following section will present the syntax and applications of typedef.

3.6.2 Syntax and applications of typedef

Figure 3.46 shows how to define a new type using typedef. In essence, typedef creates aliases for existing data types.

Syntax of type definition
typedef   originalType newType;

Figure 3.46: Syntax of typedef.

In addition to creating aliases that are intuitive and easy to remember, another use case of typedef is to simplify complex type declarations. Figure 3.47 shows two examples of typedef.

	Example 1	Example 2
Declare a new type	typedef int integer;	typedef struct student Stu;
Statement	integer x,y;	p=(struct student *)malloc(sizeof(struct student));
Equivalent statement	int x,y;	p=(Stu *)malloc(sizeof(Stu));

Figure 3.47: Examples of typedef.

In example 1, we create an alias integer for int; integer and int are equivalent types.

In example 2, we have a structure type struct student; create an alias Stu for it, so we can replace all occurrences of struct student with Stu, thus making the code easier to read.

The difference between #define and typedef is as follows: #define is a simple text replacement that happened in the preprocessing phase, while typedef enables flexible type replacement during compilation.

3.7 Summary

This chapter discusses how to describe, store, and reference a group of data that are logically correlated. Figure 3.48 shows concepts related to structures, while those of unions and enumerations are shown in Figures 3.49 and 3.50, respectively.

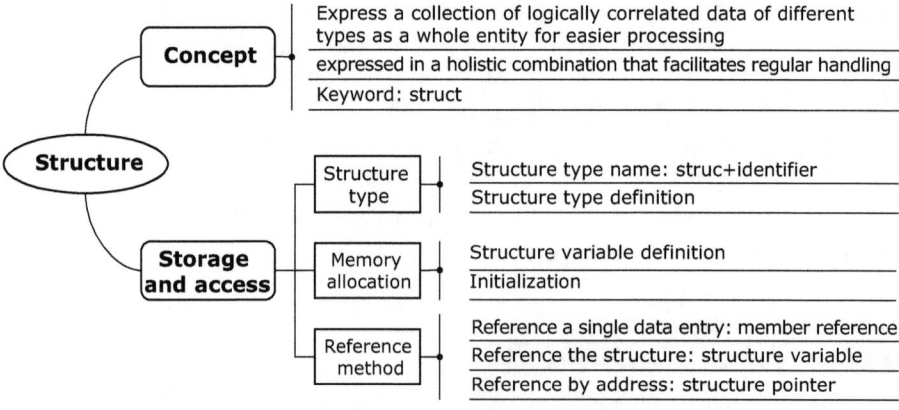

Figure 3.48: Concepts related to structures.

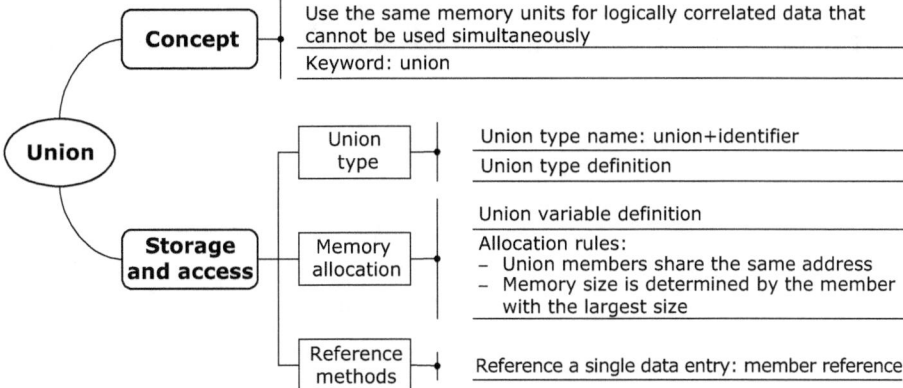

Figure 3.49: Concepts related to unions.

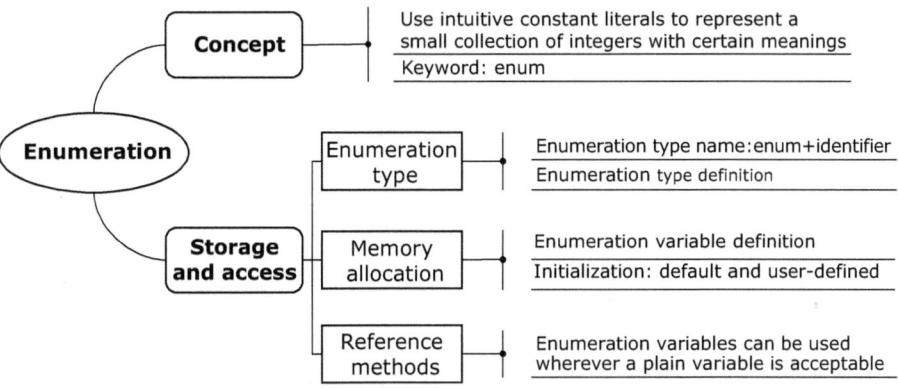

Figure 3.50: Concepts related to enumerations.

To store a group of data of different types,
We need new storage and access mechanisms other than arrays,
An aggregation of data is called a structure,
Whose size is determined by members programmers put in it.
We need to define structure variables to obtain memory space,
We can define variables, arrays, or pointers of structure types.
Members of a structure can also be accessed,
Through reference by name or by address.

Variables of different types that share the same memory space construct a union,
Whose size is determined by members programmers put in it,
A variable must be a member of the union to use the shared space,
The actual space is allocated upon definition of a union variable,

Referencing a union variable is similar to referencing a structure variable,
The valid value of the shared space depends on which member stays in it currently.

Different colors have different names,
Which correspond to abstract numbers in computers,
It is difficult to use these numbers,
So, we list color names in an enumeration to represent the integer constants,
Again, we can define enumeration types and enumeration variables,
The value of an enumeration variable has to be one of the enumeration members.

Custom types often contain many members, making it hard to use,
Type size may vary across platforms,
But some applications use fixed-length types,
Which make code porting difficult.
Hence, we rename types with typedef,
So, it suffices to modify a few places.

3.8 Exercises

3.8.1 Multiple-choice questions

1. [Array element: reference using pointers]

```
struct student
{ char name[20];
   char sex;
   int age;
} stu[3]={"Li Lin", 'M', 18, "Zhang Fun", 'M', 19, "Wang Min", 'F', 20};
struct student *p;
p=stu;
p+=2;
printf("%s, %c, %d\n", p->name, p->sex, p->age);
```

The output of the program above is ( )
A) Wang Min,F,20   B) Zhang Fun,M,19   C) Li Lin,F,19   D) Li Lin,M,18

2. [Chain structure]

```
struct sT
{ int  x; struct sT *y; } *p;
struct sT a[4]={20,a+1,15,a+2,30,a+3,17,a };
int main(void)
```

```
{ int i;
  p=a;
  for(i=1; i<=2; i++) { printf("%d,", p->x );  p=p->y; }
  return 0;
}
```

The output of the program above is ( )
A) 20,30,   B) 30,17   C) 15,30,   D) 20,15,

3. [Array element: reference using variables]
   The output of the following program is ( )

```
struct abc
{ int a, b, c; };
struct abc sum[2]={{1,2,3},{4,5,6}};
int t;
t=sum[0].a + sum[1].b;
printf("%d \n", t);
```

A) 5   B) 6   C) 7   D) 8

4. [Array elements: referencing internal elements]

```
typedef struct
{ char name[10];
  int age;
} ST;
ST stud[10]={ "Adum", 15, "Muty", 16, "Paul", 17, "Johu", 14, };
```

Which of the following is not character "u"? ( )
A) stud[3].name[3]
B) stud[2].name[2]
C) stud[1].name[1]
D) stud[0].name[3]

5. [typedef]
   Which of the following statements is wrong? ( )
   A) We can use typedef to create new types.
   B) We can use typedef to create a new name for an existing type.
   C) After defining a new type name with typedef, the original type name is still valid.
   D) We can use typedef to define aliases for existing types, but we cannot define aliases for variables.

6. [Unions]

Character "0" has decimal ASCII value 48. Suppose the 0th element of an array is stored at lower bytes and sizeof(int) is 4 bytes. What is the output of the following program? ()

union

```
{  int i[2];
     long k;
     char c[4];
}  var, *s=&var;
s->i[0]=0x39;
s->i[1]=0x38;
printf("%c\n", s->c[0]);
```

A) 39   B) 9   C) 38   D) 8

3.8.2 Fill in the tables

1.   [Operations on structure members]

Suppose we have the following structure definition. Figure out values of structure members shown in Figure 3.51 after executing the following program:

```
#include <stdio.h>
#define N 5
#define M 4
struct person
{
  int Id;
  char Name[10];
  int Score[M];//Grade
  int total;//Total grade
};
int main(void)
{
  struct person allone[N]
  ={{ 1,"mark", { 9,6,8,7 },0 },
   { 2,"bob",   { 8,6,8,5 },0 },
   { 3,"alice", { 5,9,7,8 },0 },
   { 4,"william",{ 8,9,9,9 },0 },
   { 5,"eric",  { 8,9,6,9 },0 } };
  struct person temp;
  int i, j;
  for ( i = 0; i < N; i++) //——①
  {
```

```
    allone[i].total = 0;
    for ( j = 0; j < M; j++)
    {
      allone[i].total += allone[i].Score[j];
    }
  }
  for (i =1; i < N; i++) //——②
  {
    for (j = 0; j < N-i; j++)
    {
      if (allone[j].total < allone[j+1].total)
      {
        temp = allone[j];
        allone[j] = allone[j+1];
        allone[j + 1] = temp;
      }
    }
  }
  return 0;
}
```

	i	0	1	2	3	4
After for loop ①	allone[i].id	1	2			
	allone[i].total					
After for loop ②	allone[i].id					
	allone[i].total					

Figure 3.51: Composite data: fill in the tables, question 1.

3.8.3 Programming exercises

1.   Suppose we have the following structure definition:

```
struct person
{
    char lastName[15];
    char firstName[15];
    char age[4];
}
```

Please write code that reads 10 person objects (lastName, firstName, and age) from keyboard input.

2. Please define a union to represent a point in one-dimensional space, two-dimensional space, or 3-dimensional space. The union should contain an indicator of the dimension and coordinates of the point.

3. Suppose we have the following enumeration declaration for days in a week: enum day{Sunday,Monday,Tuesday,Wednesday,Thursday,Friday,Saturday} Figure out the following values of expressions. Suppose that in each subquestion, the value of today (whose type is day) before evaluating the expression is Tuesday.

   (1) int(Monday)
   (2) int(today)
   (3) today < Tuesday
   (4) day(int(today) + 1)
   (5) Wednesday + Monday
   (6) int(today) + 1
   (7) today >= Tuesday
   (8) Wednesday + Thursday

4 Functions

4.1 Concept of functions

4.1.1 Introduction

4.1.1.1 Modularization and module reuse in practice

Case study 1 Combinatorial problems
In practice, we often need to complete a task repeatedly. For example, there are mathematical formulas for computing number of m-permutations of n elements P_m^n and number of m-combinations of n elements C_m^n, as shown in Figure 4.1. Multiple factorial computations are necessary in these formulas, which can be done by calling the factorial module multiple times in programs.

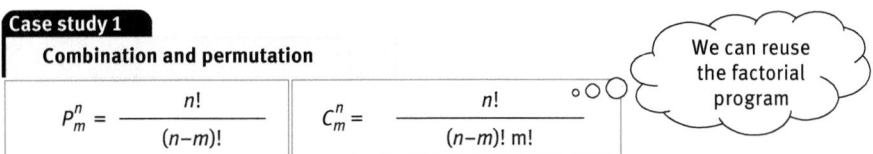

Case study 1

Combination and permutation

$$P_m^n = \frac{n!}{(n-m)!} \qquad C_m^n = \frac{n!}{(n-m)!\ m!}$$

We can reuse the factorial program

Figure 4.1: Computation of permutation and combination.

https://doi.org/10.1515/9783110692303-004

Case study 2 Scholarship application process

The university Mr. Brown works for provides scholarship to students every year. The application process is shown in Figure 4.2. Workloads of some steps in this process are so large that dedicated personnel are necessary. For example, step 3 involves computing sum in a data table; step 5 involves sorting and classifying a table; and step 6 involves searching, deletion, and insertion in a table.

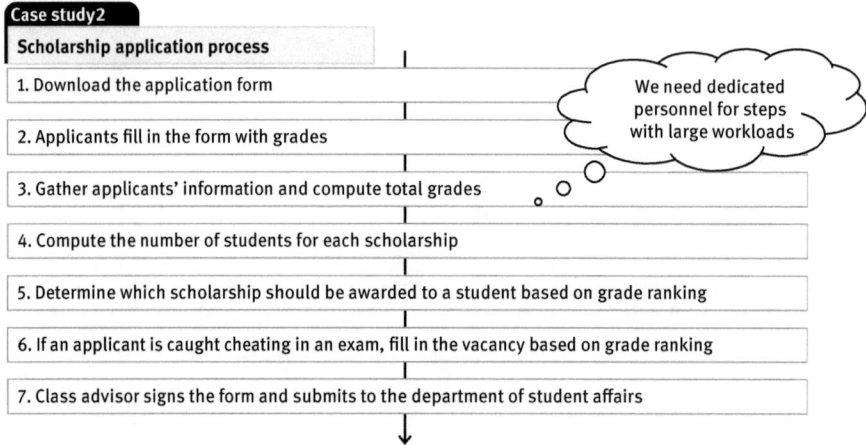

Figure 4.2: Scholarship application process.

4.1.1.2 Abstraction of practical problems: independent code modules

Usually, many issues arise when we try to solve practical problems with programs. If the scale of a problem is large and required functionalities are complex, programmers often need to work in a team to solve it. They divide the problem into modules based on functionalities so that programmers can work on different modules simultaneously. Sometimes, a function is required by most of the team, so they can implement it in an independent code module so that it can be reused. In essence, all these issues require modularization, as shown in Figure 4.3.

Problems	Strategy	Solution
Scale of problems is large Functionality is complex	Teamwork	Divide into modules based on functionality – Write in modules – Test in modules
We want to reuse programs	Reuse	Build independent code modules

Figure 4.3: Independent modules.

4.1.2 Concept of modules

4.1.2.1 Coordination problems in teamwork

Before discussing the modularization mechanism in programs, let us examine and analyze how humans solve real-life problems. Figure 4.4 shows critical steps that require cooperation in the scholarship application process we just saw. What are the differences between doing the work on one's own and doing the work in a team? Figure 4.5 compares these two ways from the perspective of workload, nature of the work, and necessity of communication.

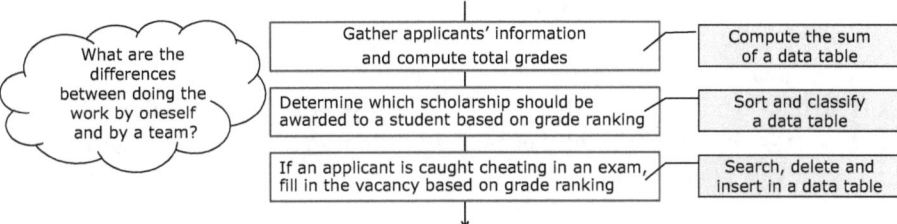

Figure 4.4: Work that requires cooperation in the scholarship application process.

	Individual	Team
Workload	Large	Small
Work nature	Composite	Single
Communication	Not necessary	Necessary

Figure 4.5: Analysis of individual work and teamwork.

The workload is heavy for one person but small for a team. Multiple skills are necessary for a single individual to complete all the work, while one skill may suffice for a person in a team to complete the task assigned to him/her. There is no need to communicate if a single person does the work, but communication is of great significance in teamwork because the output of one step is often the input of the next.

4.1.2.2 Coordination problems in modularization of programs

Similarly, what issues exist in the modularization of programs?

Think and discuss Issues of modularization of programs

1. What are the differences between using one segment of code and using multiple segments of code to solve a problem?
2. What is the key to using multiple child programs to complete one task?

Discussion: There is no difference in workload or complexity of these two ways. However, using several child programs require information transfer, which is precisely the key we are looking for. Hence, programming languages must provide such mechanisms.

4.1.2.3 Concept of modules

Based on the discussion above, we can summarize what is necessary for independent code modules.

We call collections of statements that have its own name and can complete specific tasks independently "modules" in programming, as shown in Figure 4.6. A module consists of the implementation and an interface. Interfaces are created to hide a module's implementation and data from outside of the module. Communication with external objects must be done through information interfaces. The interface of a module describes how other modules or programs should use it. Input/output information is also part of an interface.

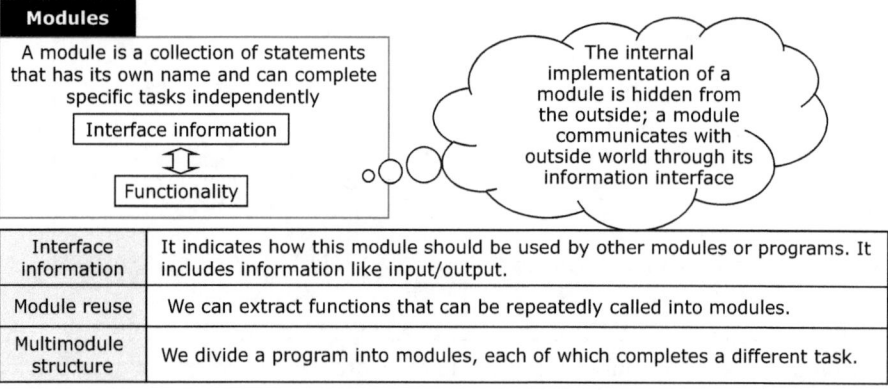

Interface information	It indicates how this module should be used by other modules or programs. It includes information like input/output.
Module reuse	We can extract functions that can be repeatedly called into modules.
Multimodule structure	We divide a program into modules, each of which completes a different task.

Figure 4.6: Concepts related to modules.

There are other concepts related to modules, such as module reusing and multimodule structures.

The word "module" has many aliases, such as function or child program. C uses "function" to describe modules, as shown in Figure 4.7. We use the word "module" in structured analysis and design; and it becomes "class" in object-oriented analysis and design; the term used in component-based development is "component."

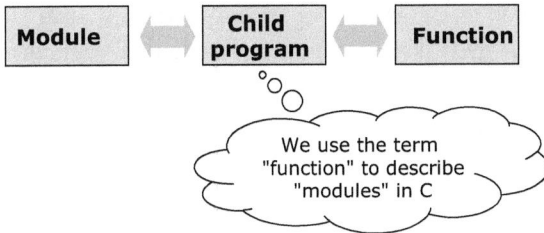

Figure 4.7: Aliases of modules.

Modularized program design has the following features:
(1) Modules are independent of each other. Each module has its functionality. Programs using modules have more lucid logic and are easier to write and maintain.
(2) It is easier to design programs, so the development cycle is shortened.
(3) Modules are more robust.
(4) Programmers no longer have to reinvent the wheel.
(5) It is easier to maintain existing code and write new code.

4.2 Function form design

4.2.1 Methods of communication between modules

Mr. Brown's university is going to hold a commencement ceremony. The rostrum is going to be built by the logistics department. Figure 4.8 shows the steps in the building

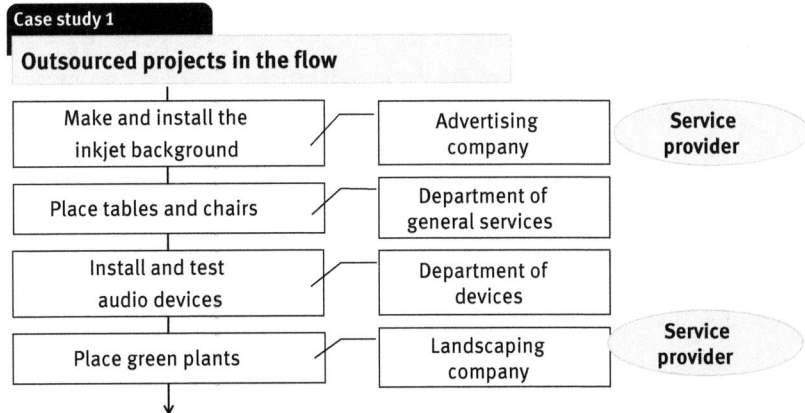

Figure 4.8: Process of building rostrum of the commencement ceremony.

process. Some steps can be outsourced to professional companies, such as advertising companies or landscaping companies, to accelerate the process and guarantee the quality.

Think and discuss Coordination methods in outsourced projects
Discussion: There are two methods of communication in an outsourced project, as shown in Figure 4.9.
(1)   The outsourcer coordinates between service providers (e.g., rostrum building process).
(2)   Service providers communicate with each other (e.g., scholarship application process).

Both methods are feasible in programming design patterns. If modules are executed in order, then the program is procedure oriented; if a module would not be executed until certain events happen, then the program is object oriented. Readers can refer to Appendix B for more details on this topic. Functions in C use the first method mentioned earlier, where "the outsourcer coordinates between service providers."

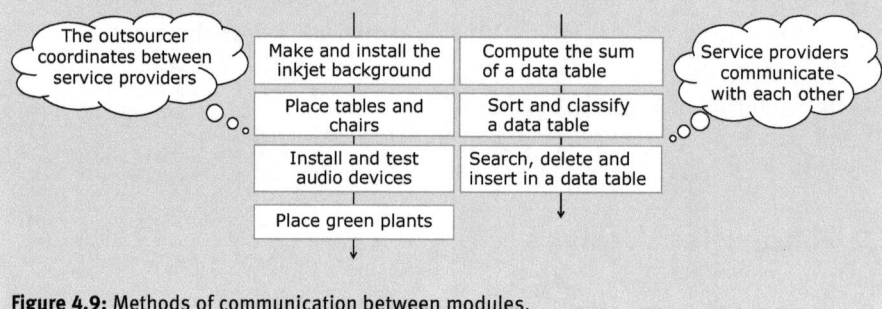

Figure 4.9: Methods of communication between modules.

4.2.2 Function form design

4.2.2.1 Analysis of outsourcing structure
Let us take the advertising company as an example to analyze the structure of outsourcing.

In the "background production" project, the advertising company is the service provider, and the university is the outsourcer, as shown in Figure 4.10. The advertising company needs to make a statement about what they can do, including materials they use, specifications of the materials, rendering preview, and service price. This statement can be considered as a definition of "production." The definition does not produce actual products. The advertising company will not start "producing" based on the definition until the outsourcer provides image assets, size of background, and quote. In other words, it is the outsourcer that "drives the production."

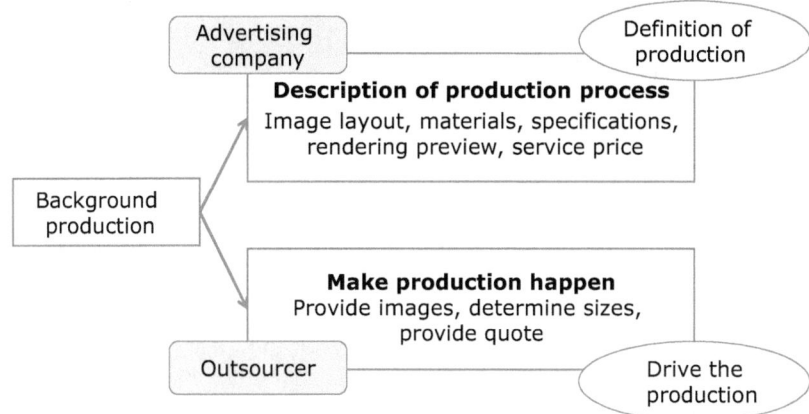

Figure 4.10: Analysis of the outsourcing structure.

4.2.2.2 Abstraction of outsourcing structure

Figure 4.11 illustrates a further abstraction of the outsourcing structure. We can consider outsourcing as a "manufacture" process. There are three critical elements in manufacturing: input, output, and processing. Input is material used in the process. The output is the final product. Processing refers to procedures in the manufacturing process. To start manufacture, users need to provide materials required by the manufacturer. In programs, these materials are simply data.

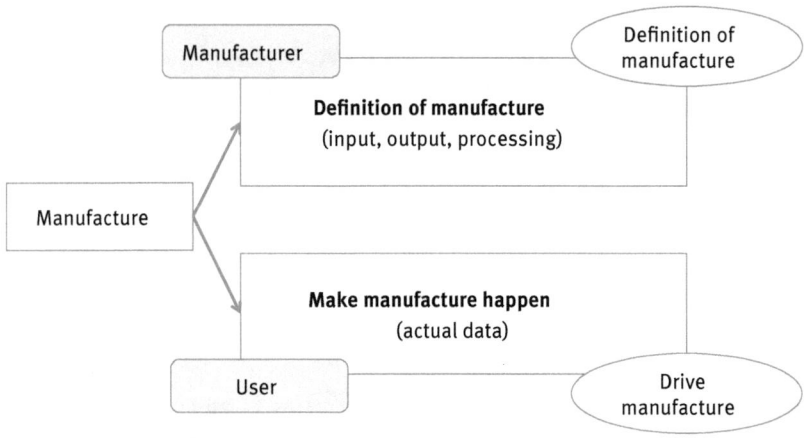

Figure 4.11: Abstraction of the outsourcing structure.

4.2.2.3 Function form design

As independent code segments, functions are similar to outsourcing projects, as shown in Figure 4.12. The process of writing a function defines its functionality,

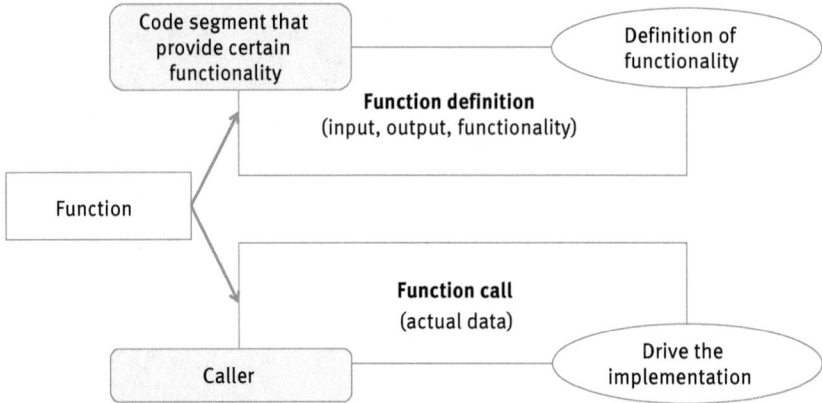

Figure 4.12: Function form design.

which includes input, output, and processing. Users of functions are referred to as "callers" in programming languages. Callers provide functions with actual data so that they can complete specific tasks.

4.2.2.4 Information transmission mechanism design

In practice, we can deliver raw materials to manufacturers through express or Internet. When manufacture is completed, the manufacturer can send the product back in a way suitable to users. In programs, however, data processing is done entirely in computers. Thus, the information transmission mechanism needs to conform to the characteristics of computers.

When designing mechanisms of functions, we need to determine how the caller sends data to the function and how the function sends results back to the caller, as shown in Figure 4.13.

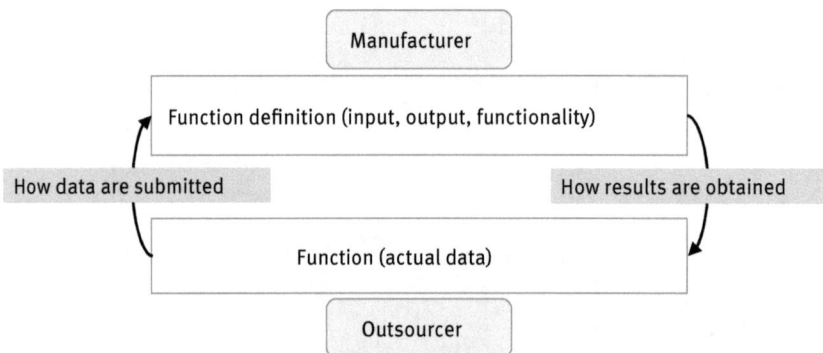

Figure 4.13: Information transmission mechanism design.

One of the methods of information transmission between modules is using software interfaces, as shown in Figure 4.14. For "manufacturers," namely those who define functions, they should consider the following issues: interfaces of receiving information, code implementation, and approach of result submission. To describe the manufacturing project, we need to determine a name for it, which is the function name. For "outsourcers," namely function callers, they need to consider the following issues: interfaces of submitting information and approach of receiving results.

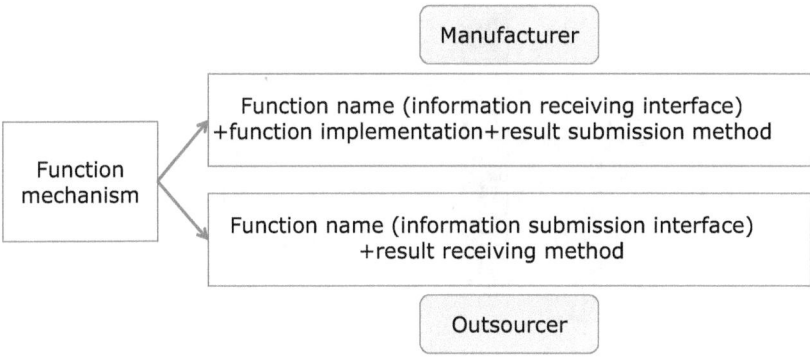

Figure 4.14: Information transmission design in functions.

4.2.2.5 Three syntaxes related to functions

C offers three syntaxes related to functions: function definition (the "manufacturer"), function call (the "outsourcer"), and function declaration (or function prototype), which briefly describes a function. Figure 4.15 shows these syntaxes.

In a function definition, the information interface is implemented as a "parameter list," and implementation is done by declarations and statements between the curly brackets. The type of function result is determined by "function type." In a function call, the interface of submitting the information is implemented as an "argument list." C provides two "approaches of result submission:" one is to submit through information interface and the other is to use return statements. The function call syntax does not explicitly show how the caller receives results. We shall cover result submission and receiving in detail later.

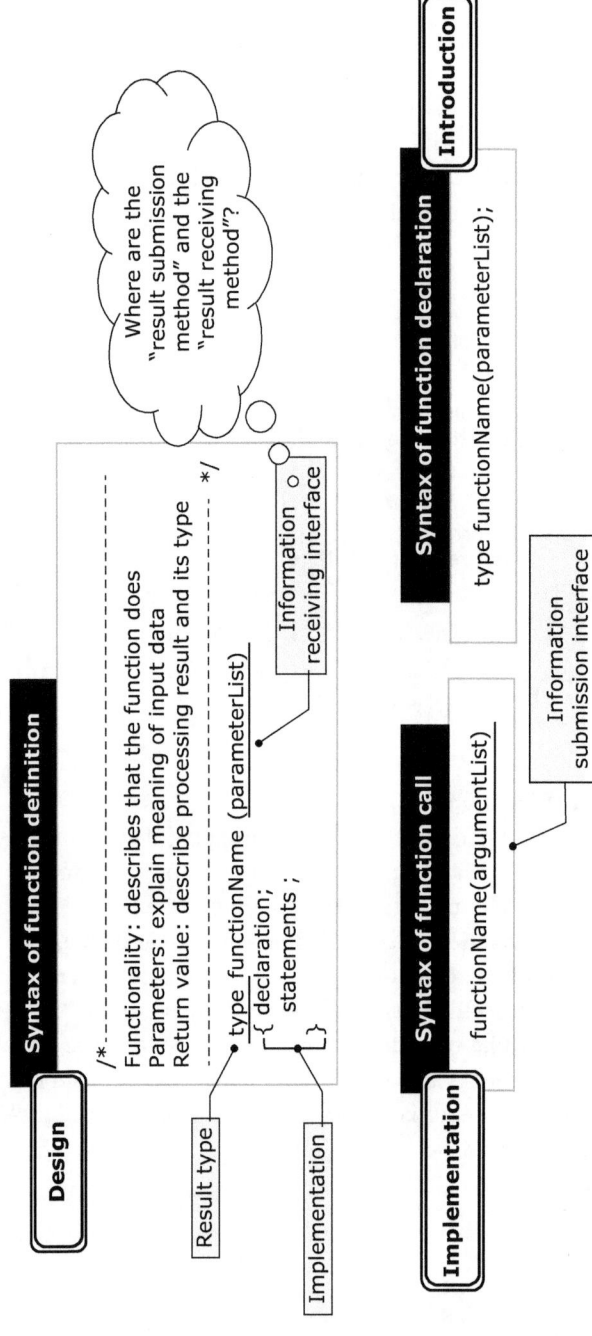

Figure 4.15: Three syntaxes related to functions.

Knowledge ABC Function declaration and where to write them
C is an old language, so its grammar has been revised continuously. There used to be few restrictions on the order of function declaration, definition, and call and necessity of function declaration. Different compilers also followed different rules on them. However, the latest C standards (such as C99 and C11) require that a function must be declared before being called. This requirement helps compilers find out errors of argument types and numbers in function calls. In most cases, function declarations should be written at the beginning of code (usually after preprocessing directives) and outside of function definitions. Like before, the new standards put no limits on the location of function definitions and function calls.

4.3 Design of information transmission mechanism between functions

4.3.1 Characteristics of information transmission between functions

4.3.1.1 Classification of data transmitted between functions

We have concluded in the introduction of functions that the key to completing a task using multiple child programs is the information communication between them. Programming languages should provide mechanisms for such communication.

In the scholarship example, fundamental steps that require cooperation are computing sum, classification, and applicant substitution. We can implement them in three functions, in which operations like addition, sorting, classification, searching, deletion, and insertion are involved.

Figure 4.16 lists input/output information needed in these steps. These data are either single data entries or groups of correlated data. The sum function reads raw data table and outputs the data table with an extra column of the total score. The classification function reads raw data table and classification arguments, and outputs the number of students of each level and their names.

Function	Input	Output	Notes
Sum	– Data table	Data table	The output data table contains total grades
Classification	– Data table – Classification parameters	Number of students of each scholarship and student names	Sorting before classification
Substitution	– Sorted and classified data table – Name of students to be deleted from the table	Number of students of each scholarship and student names	There are data of individuals and data of a group of students

Figure 4.16: Analysis of data in critical steps of the scholarship application process.

! **Think and discuss** In the process of transmitting data to function through its interface, what are characteristics of the data and how are they transmitted?

Discussion: As shown in Figure 4.17, issues related to data transmission are data type and data size. Since the nature of types is the size of the memory space used, type issues are necessarily size issues. We shall discuss the later below.

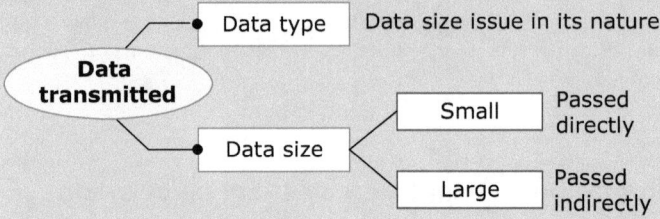

Figure 4.17: Characteristics of information transmitted between functions.

In real life, we can send items to others directly or indirectly. For example, a mail carrier can send parcels to recipients directly or put them in a self-service parcel pick-up machine.

We can use these methods in information transmission between functions as well. In programs, a small amount of data are often passed to functions directly. In contrast, a large amount of data are often passed to functions indirectly by providing the beginning address of the data so that functions can fetch them on their own.

4.3.1.2 Expressions of data transmitted between functions

Data to be processed have different names if we consider them from different perspectives. From the function caller (outsourcer) standpoint, the data to be processed are called "arguments" in C; from the function definition (manufacturer) standpoint, the data it receives through software interface are called "parameters," as shown in Figure 4.18. As for how the final product is transmitted to users, we shall cover the topic later.

4.3.2 Information transmission between functions: submission and receiving of data

As we have discussed earlier, C processes data in different ways, according to the size of data passed to a function.

4.3.2.1 Submission of small amount of data

We shall start from a small amount of data.

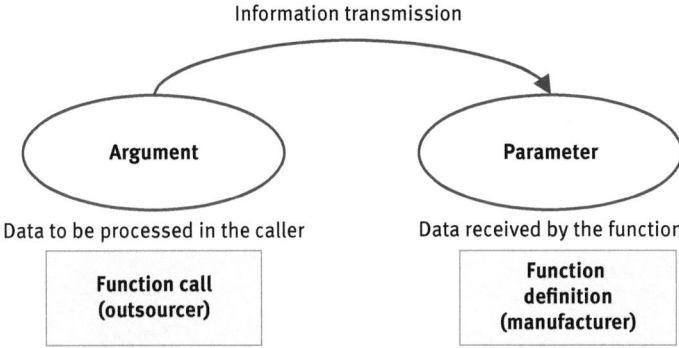

Figure 4.18: Terms for data used in functions.

The memory space allocated to actual values that a function caller needs to process is called "argument space." The system copies the actual data and sends it to the function called, as shown in Figure 4.19. In other words, the function receives a copy of actual values. We can imagine the process as sending copies of assets to advertisement companies for printing. The memory space allocated to these copies is called "parameter space."

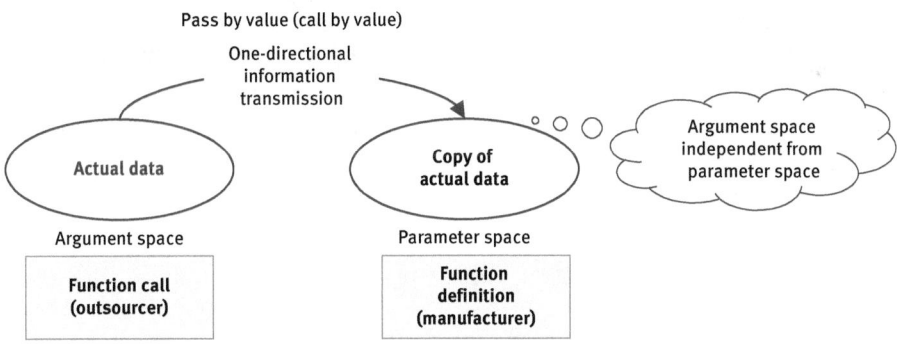

Figure 4.19: Small amount of data: passed directly.

Argument space is independent of parameter space. Thus, updates on data in parameter space will not change data in argument space. Such information transmission is single directional. Because the data transmitted are values, we call the process "pass by value." Such way of calling functions is called "call by value."

4.3.2.2 Submission of a large amount of data
Now we are going to study the case of a large amount of data.

If the size of the data is large, the cost of passing copies is also high, which affects communication efficiency. In this case, the function caller can pass the beginning address of the data to the "manufacturer," because a large amount of data are usually stored continuously in memory. The "manufacturer" then fetches data from the specified address, as shown in Figure 4.20.

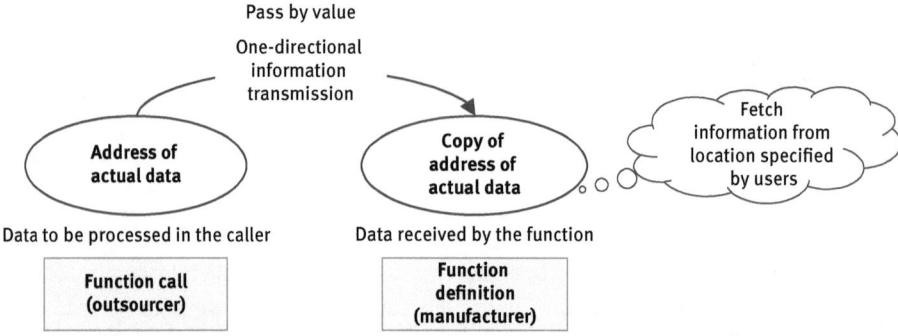

Figure 4.20: Large amount of data: submit the address of data.

Note that the information passed is a copy of the "address of actual data." It is similar to data maintenance of library servers: service providers can operate servers remotely as long as they know IP addresses and passwords.

Service providers can also carry out maintenance on-site. Similarly, the "manufacturer" can also head to the address of data and process them directly, as shown in Figure 4.21. This is "pass by reference" in C. Function callers can pass the beginning address of data to the "manufacturer" so that it can process data at the address directly.

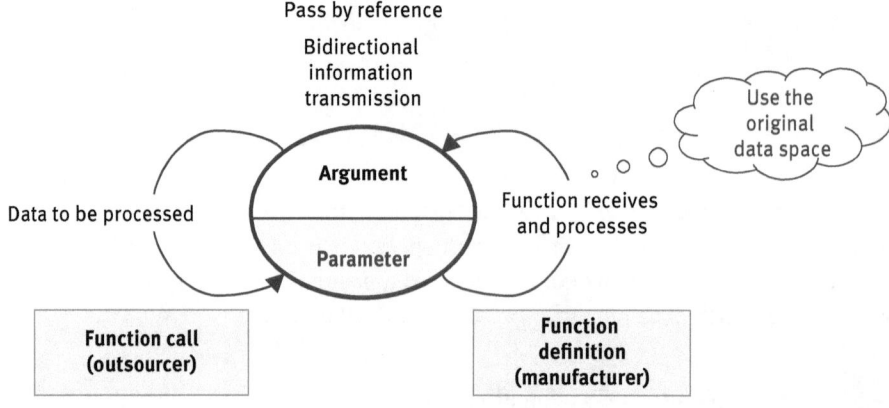

Figure 4.21: Large amount of data: processed on-site.

Note that argument space and parameter space is the same in "pass by reference."

We have just seen that function callers can pass data to functions by value or by reference. In fact, how function results are received are related to how data are submitted.

4.3.3 Receiving of function results

4.3.3.1 Receiving function results in pass by value

In the case of pass by value, C provides two ways of submitting results, as shown in Figure 4.22. The first is using a return statement to pass a single result. The second way works for address parameters. Function callers can find results at this address. In this case, many results can be passed.

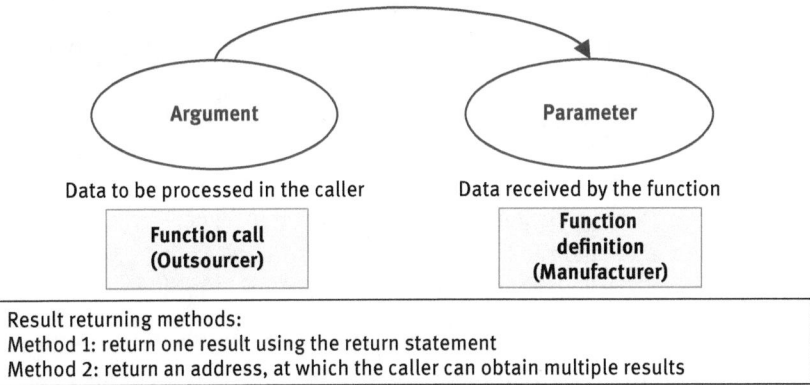

Figure 4.22: Submission and receiving of result in pass by value.

4.3.3.2 Receiving function results in pass by reference

In the case of pass by reference, function callers can obtain multiple results in the shared data space, as shown in Figure 4.23.

4.4 Overall function design

4.4.1 Key elements of function design

4.4.1.1 Key elements of functions

In the discussion of function form design and information transmission mechanism design, we compared function modules to "manufacturers" in real life. Figure 4.24 summarizes the relations we found in the comparison. There are three key elements of functions: input, output, and processing.

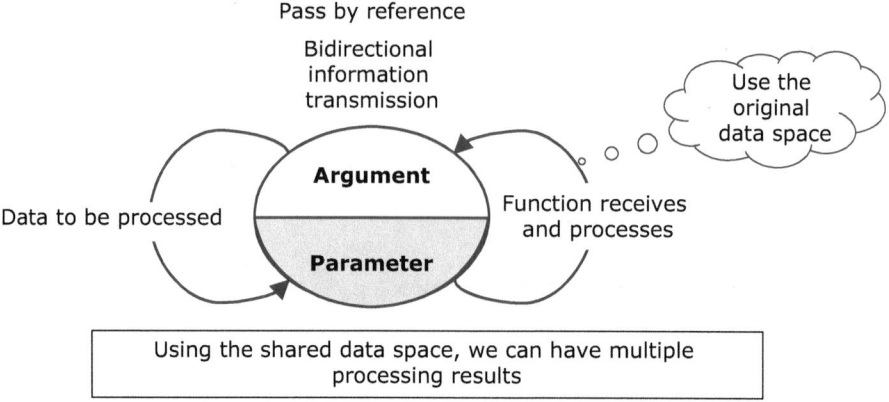

Figure 4.23: Submission and receiving of result in pass by reference.

A function name briefly describes the processing, while the function body implements the processing.

Data receiving and result submission of a function are done through information interfaces. The input information interface is implemented as the parameter list. The information passed can be either values or addresses. Results can be output in two ways: using a return statement or putting them at a specified address for callers to access. Again, results can be either values or addresses.

4.4.1.2 Relations between function syntax and key elements of function design

As shown in Figure 4.25, input information determines the parameter list, while output information determines the function type.

A function in C consists of a function header and a function body. The function header describes the structure of a function, while the function body implements its functionality. As such, input, output, and processing of a function determines its structure.

4.4.2 Summarization of information transmission between functions

4.4.2.1 Direction 1: from caller to function

In C programs, information is passed from arguments (user data) to parameters (manufacturer data) in two ways: pass by value and pass by reference, as shown in Figure 4.26. Arguments and parameters are stored in separate space in the case of pass by value, while they share the same space in the case of pass by reference. There is another way in C called simulated pass by reference, in which data passed is address, but parameters and arguments have different spaces. In high-level languages,

Manufacture process	Data in necessary information			Functionality	Data in submitted results		
	Type of information	Quantity			Type of data	Quantity	
	Value, address	≥ 0			Value, address	≥ 0	
Function name	Input information (interface information)			Function body implementation	Output information (interface information)		
	Input information parameter list: (type variable 1, type variable 2,....)				Single: return (value) Multiple: store at specified addresses (1) return (address) (2) specify the address in parameter list		

Figure 4.24: Key elements of function design.

Functionality	Input information	Output information
Function name briefly describes the functionality	The characteristics and quantity of input information determine the parameter list	The type of output information determines the function type
Function name	**Parameter list**	**Function type**

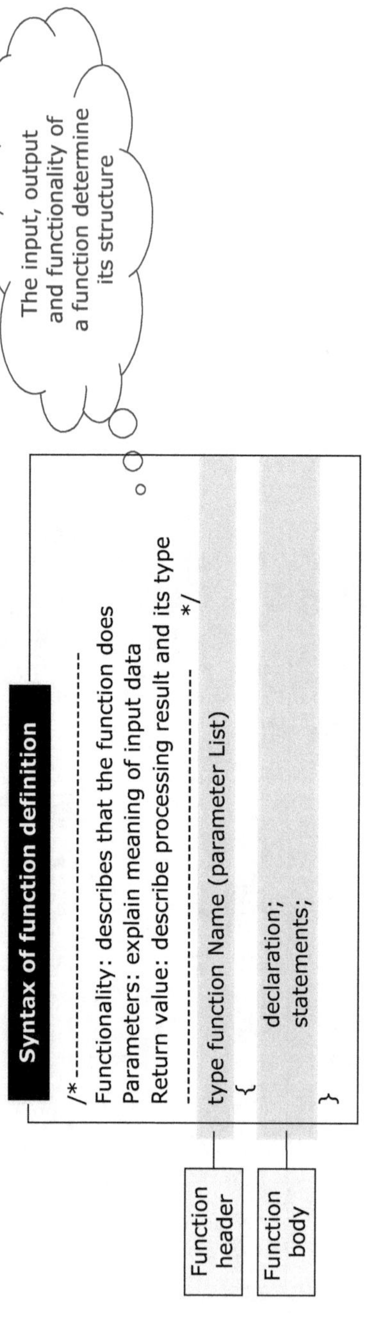

The input, output and functionality of a function determine its structure

Syntax of function definition

```
/*-------------------------------------------
Functionality: describes that the function does
Parameters: explain meaning of input data
Return value: describe processing result and its type
--------------------------------------------- */
type function Name (parameter List)
{
    declaration;
    statements;
}
```

Function header

Function body

Figure 4.25: Key elements of function design and their expressions.

Direction	Method	Allocation of memory space	Type of call
Argument->parameter	Pass by value	Arguments and parameters have separate memory units	Call by value
	Simulated pass by reference		
	Pass by reference	Arguments and parameters share memory units	Call by reference

Figure 4.26: Data transmission from the caller to function.

a function call via pass by value is called a "call by value," while one via pass by reference is called a "call by reference."

4.4.2.2 Direction 2: from function to caller

Figure 4.27 shows ways of passing results. If the result is a single value, we can use a return statement to pass it. If the result contains multiple values, we can return an address. Additionally, if a parameter is an address, we can store these values at the address passed in by the caller so that the caller can fetch them. A function may return nothing in special cases.

Processing result	Method
Single value	return(value)
Multiple values	return(address)
	Parameter is an address

Special case: no return value

Figure 4.27: Data transmission from function to caller.

4.4.3 Function call

4.4.3.1 Execution and calling order of functions

In the rostrum building example in Section 4.2, two steps are outsourced to service providers. Figure 4.28 lists substeps of these two steps.

The execution order of functions is similar to the earlier process. The main function is the outsourcer, while child functions are service providers of steps in the process. During execution, a program always starts from the main function. When a child function is called, the program enters the child function and returns to the main function after the child function terminates. Figure 4.29 illustrates the entire process.

4.4.3.2 Nested call of functions

After the midterm exam, Mr. Brown would like to know the highest score, the lowest score, and the difference between the two, so he asked his class representative A to compute these values.

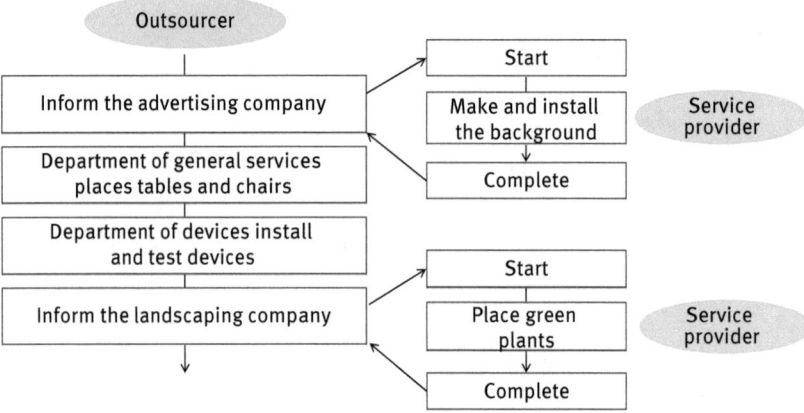

Figure 4.28: Rostrum building process with outsourced steps.

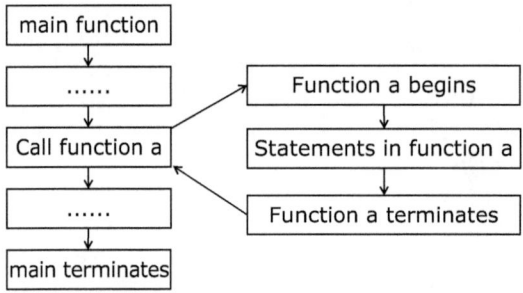

Figure 4.29: Execution order of functions.

A completed the task quickly and reported the result to him. Then Mr. Brown asked, "If we simulate this process with a program and you are asked to implement the child functions, how are you going to do that?"

"That's simple," answered A, "I'll write two functions. The max function computes the highest score, and the min function computes the lowest score. The main function can obtain the highest and the lowest scores by calling them, and then compute the difference."

Mr. Brown smiled and asked, "Is that a complete simulation?" A thought for a while and responded, "No, I should have written another function for difference computation." "How does this function work then?" Mr. Brown followed up.

A said, "Let the function be dif, then the execution order of these functions is as shown in Figure 4.30. We have learned nested if and nest loops, can we call this 'nested function call'?" "Of course," Mr. Brown commended, "We do use this term in C."

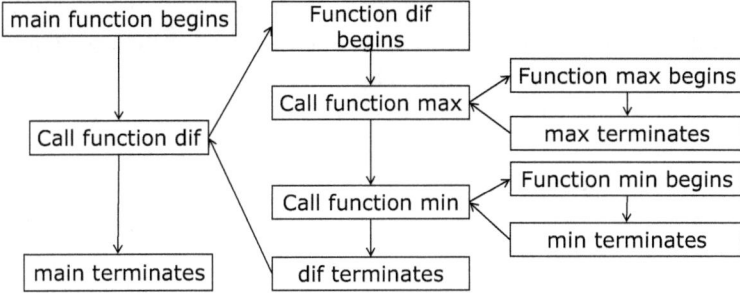

Figure 4.30: Execution order of nested function call.

In C programs, a function can call another function, and the function being called can further call other functions, resulting in a nested function call. We may have arbitrary layers of nested calls and complete sophisticated tasks with them.

All C programs are constructed by functions, each of which is independently defined. That is, one cannot define another function in the definition of a function.

4.4.3.3 Correspondence between parameters and arguments
Information to be received is defined in the parameter list of a function definition. When calling this function, one needs to put data to be processed into the argument list. The parameter list contains definitions of variables, while the argument list contains references of variables, as shown in Figure 4.31.

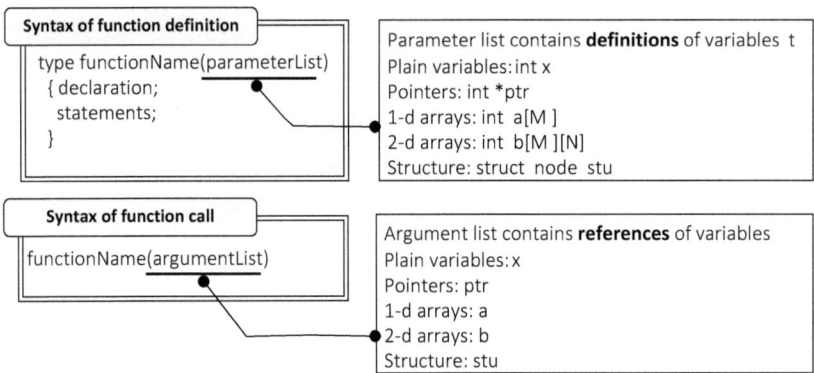

Figure 4.31: Correspondence between parameters and arguments.

When using functions, one should always use the correct syntax. One of the common mistakes beginners make is using the wrong arguments. Programs with such mistakes cannot be compiled. Furthermore, it is often hard for them to realize the mistake.

(1) Parameter list: In a function definition, the definitions of parameters are listed in the parameter list.

(2) Argument list: In a function call, references of arguments are listed in the argument list; in the case of arrays, we simply use array names in the argument list.

4.4.3.4 Syntax of function call

Based on whether a computation result exists, functions in C are classified into value-returning functions and nonvalue-returning functions. A value-returning function has a computation result and uses its type as the function type; a nonvalue-returning function processes the data with no explicit computation result. For example, a sorting function sorts data but does not compute a result.

A variable is necessary to store the result returned by value-returning functions, while it is not for nonvalue-returning functions. Figure 4.32 shows syntaxes of calling both kinds of functions.

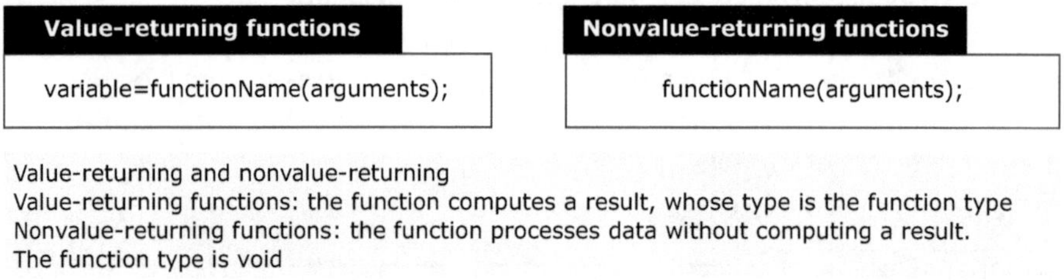

Value-returning functions	Nonvalue-returning functions
variable=functionName(arguments);	functionName(arguments);

Value-returning and nonvalue-returning
Value-returning functions: the function computes a result, whose type is the function type
Nonvalue-returning functions: the function processes data without computing a result.
The function type is void

Figure 4.32: Syntax of a function call.

4.5 Examples of function design

We have introduced the concept of functions, information transmission mechanism between functions, and key elements of function design in previous sections. Now we are going to study some examples of function design.

4.5.1 Call by value

Example 4.1 Finding maximum of three numbers
1. Function structure design
In function structure design, we extract input, output, and processing from the problem description. This example requires us to find the maximum using a function max. The input of this

function is three integers, which determines the parameter list. The output of the function is the computed maximum, which is an integer, so the function should be int as well. Figure 4.33 summarizes key elements of the function.

Functionality	Input information	Output information
max	int a,b,c	int value
Function name	**Parameter list**	**Function type**

Figure 4.33: Key elements of function max.

2. Comparison of using main function and using child function
We shall implement the maximum finding code in the main function and in a child function, and then compare the implementations, as shown in Figure 4.34. Values of a, b, and c in the main function are obtained from keyboard input, while they are obtained from the interface, namely parameter list, in the child function. The maximum found in the main function is directly displayed onto the screen, while that of the child function is returned to its caller through a return statement.

Through the comparison, we can conclude that although the two implementations differ in input and output, statements used to find the maximum are exactly the same.

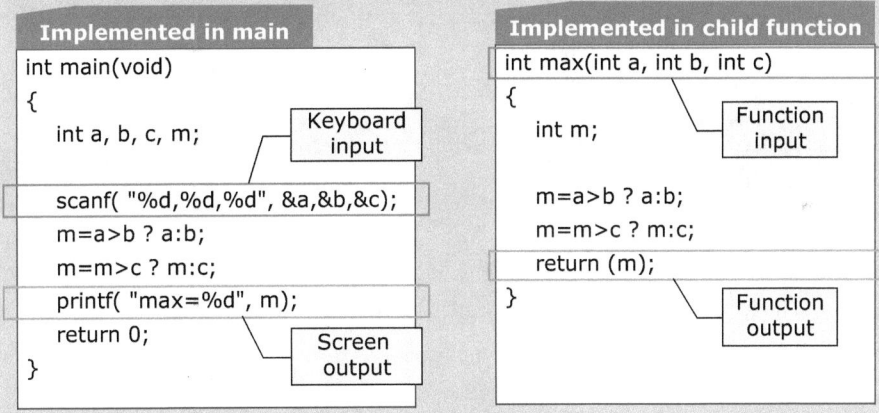

Figure 4.34: Different implementations of the max function.

3. Calling child function
A child function must be called to complete its functionality. The caller can be either the main function or other child functions. Figure 4.35 shows the program that finds the maximum using a child function. Note how the child function is declared, defined, and called, and the order in which these three constructions appear in the program. Similar to variables, a function must be declared or defined before being called.

Line 2 declares the child function max. The line is also called the header of function max.

The definition of max is between lines 6 and 12.

On line 16, the three inputs are read from the keyboard in the main function.

On line 17, the max function is called. Because max is a value-returning function, the result is stored in an integer variable x.

```
01 #include <stdio.h>
02 int max(int a, int b,  int c);              //Declare function max
03 /*------------------------------------------ -----------
04   Find the maximum among numbers a, b and c
05 ------------------------------------------ ------------------*/
06 int max(int a, int b,  int c )               //Define function max
07 {
08     int m;
09     m=a>b ? a:b;
10     m=m>c ? m:c;
11     return (m);
12 }
13 int main(void)
14 {
15     int a, b, c, x;
16     scanf("%d,%d,%d", &a, &b, &c);
17     x=max(a,b,c);                            //Call function max
18     printf("max=%d", x);
19     return 0;
20 }
```

Figure 4.35: Relation between max and main.

4. Debugging

We can study how child functions are called using a debugger.

Before debugging, we need to determine what we would like to inspect and act accordingly. Issues we are going to investigate related to call by value are shown in Figure 4.36.

Debugging plan

– Are addresses of arguments and parameters the same?
– How are parameters and arguments passed?
– Is it easy to debug if parameters and arguments have the same names?

Figure 4.36: Debugging plan of maximum finding program.

The input parameters of the main function and max function are a, b, and c. We can make a table to record their values during debugging and then analyze these values. Figure 4.37 shows the completed table with "Address" values obtained from the debugger. The debugging process is shown in Figure 4.38, in which the image on the left shows the Watch window before max is called, and the image on the right shows the Watch window after entering the max function. Command of stepwise tracing has been introduced in chapter "Execution of Programs." With values and addresses of a, b, and c displayed in the Watch window, we can complete the table in Figure 4.37. Note that the variable addresses may vary after each linking and compilation.

Arguments in main function			Parameters in child function		
Variable	Address	Value	Variable	Address	Value
a	0x0018ff44	2	a	0x0018fee0	2
b	0x0018ff40	3	b	0x0018fee4	3
c	0x0018ff3c	6	c	0x0018fee8	6
x=max(a,b,c)			int max(int a, int b, int c)		

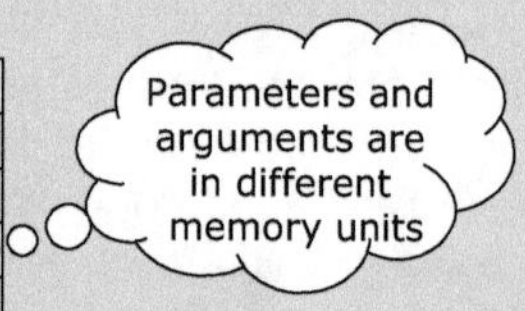

Parameters and arguments are in different memory units

Figure 4.37: Debugger data of maximum finding program.

Evidently, addresses of arguments a, b, and c are different from addresses of parameters a, b, and c. This indicates that they are stored in different spaces. It also proves that the values of arguments are copied into parameters.

Although arguments and parameters in this program have the same name, they are stored at different addresses, so they are essentially different variables. Using the same names for arguments and parameters can easily confuse programmers when debugging, so it is recommended to use different names for them.

Figure 4.38: Debugging process of maximum finding program using the same name for arguments and parameters.

We shall rename the arguments as d, e, and f, and repeat the debugging process.

5. Debugging the new implementation

Figure 4.39 shows the values of parameters and arguments used in the main function and max function when the program first enters the main function. Readers may have noticed values of variables d, e, and f:

CXX0069: Error: variable needs stack frame

This error occurs because stack memory space has not been allocated to these variables. Please refer to the chapter "Execution of Programs" for the concept of a stack.

```
#include <stdio.h>
int max( int a, int b,  int c); //
/*------------------------------
    在三个数a、b、c中，找其中的最大值
-------------------------------
int max(   int a, int b,   int c )
{
    int m;
    m=a>b ? a:b;
    m=m>c ? m:c;
    return (m);
}
int main(void)
{
    int d, e, f, x;
    scanf( "%d,%d,%d", &d, &e, &f);
    x=max (d, e, f);
    printf( "max=%d", x);
    return 0;
}
```

Watch	
Name	Value
d	CXX0069: Error: variable needs stack frame
e	CXX0069: Error: variable needs stack frame
f	CXX0069: Error: variable needs stack frame
&d	CXX0069: Error: variable needs stack frame
&e	CXX0069: Error: variable needs stack frame
&f	CXX0069: Error: variable needs stack frame
x	CXX0069: Error: variable needs stack frame
a	CXX0017: Error: symbol "a" not found
b	CXX0017: Error: symbol "b" not found
c	CXX0017: Error: symbol "c" not found
&a	CXX0017: Error: symbol "a" not found
&b	CXX0017: Error: symbol "b" not found
&c	CXX0017: Error: symbol "c" not found

Watch1 / Watch2 \ Watch3 \ Watch4 /

Figure 4.39: Debugging process of maximum finding program 1.

> ℹ️ **Knowledge ABC** Stack frame
>
> Stack frames are also called "activation records." They are a data structure used by compilers to implement function calls. Logically, a stack frame is an environment in which a function is executed. It contains all data related to a function call: parameters, local variables, return address, copies of register values that need to be restored, and so forth. Upon a function call, a frame is pushed onto the stack. After the function terminates, the frame is popped from the stack.

If we step forward in the main function, variables in the main function will obtain memory space and addresses as shown in Figure 4.40. At this moment, their values are still random numbers. Values of variables in child function max are "not found" at this moment. This is due to the masking mechanism of modules, which prevents a function from accessing data inside other functions.

Figure 4.41 shows the state after executing scanf and before calling max. Now values of d, e, and f are 2, 3, and 6, respectively. The value of x is still a random number.

Figure 4.40: Debugging process of maximum finding program 2.

By pressing F11, we step into child function max. As shown in Figure 4.42, variables in the main function are now invisible, while variables in max become visible. We can see that parameters a, b, and c have obtained values of arguments d, e, and f, but their addresses are different from those of d, e, and f. The value of m is a random number at this moment.

After max function terminates, result 6 is stored into m, as shown in Figure 4.43.

The program then steps out of max and returns to main, as shown in Figure 4.44. We can see that the value of x has become 6.

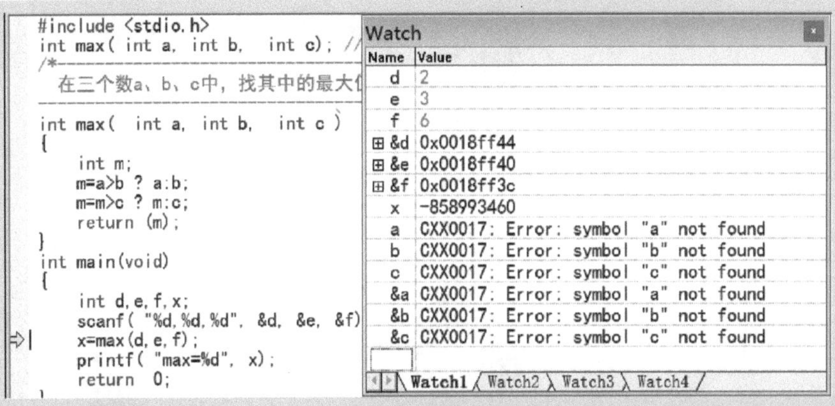

```
#include <stdio.h>
int max( int a, int b,  int c); //
/*─────────────────────────────
   在三个数a、b、c中，找其中的最大值
─────────────────────────────
int max(  int a, int b,   int c )
{
    int m;
    m=a>b ? a:b;
    m=m>c ? m:c;
    return (m);
}
int main(void)
{
    int d,e,f,x;
    scanf( "%d,%d,%d", &d, &e, &f)
⇨|  x=max(d,e,f);
    printf( "max=%d", x);
    return 0;
```

Watch	
Name	**Value**
d	2
e	3
f	6
⊞ &d	0x0018ff44
⊞ &e	0x0018ff40
⊞ &f	0x0018ff3c
x	-858993460
a	CXX0017: Error: symbol "a" not found
b	CXX0017: Error: symbol "b" not found
c	CXX0017: Error: symbol "c" not found
&a	CXX0017: Error: symbol "a" not found
&b	CXX0017: Error: symbol "b" not found
&c	CXX0017: Error: symbol "c" not found

◀▶\ **Watch1** ⟨ Watch2 ⟩ Watch3 ⟩ Watch4 ⟩

Figure 4.41: Debugging process of maximum finding program 3.

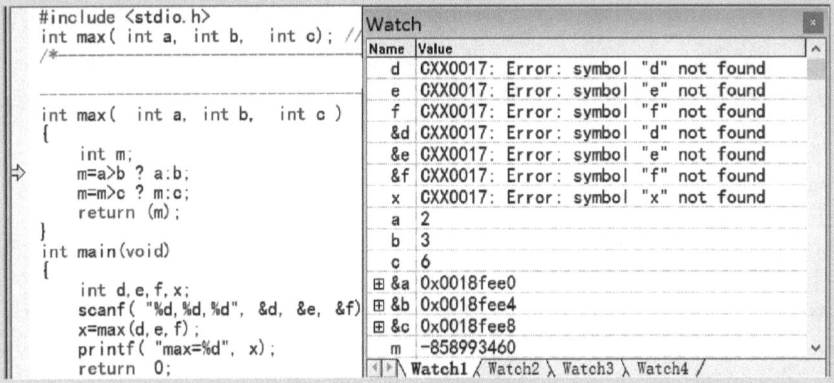

```
#include <stdio.h>
int max( int a, int b,  int c); //
/*─────────────────────────────
─────────────────────────────
int max(  int a, int b,   int c )
{
    int m;
⇨   m=a>b ? a:b;
    m=m>c ? m:c;
    return (m);
}
int main(void)
{
    int d,e,f,x;
    scanf( "%d,%d,%d", &d, &e, &f)
    x=max(d,e,f);
    printf( "max=%d", x);
    return 0;
```

Watch	
Name	**Value**
d	CXX0017: Error: symbol "d" not found
e	CXX0017: Error: symbol "e" not found
f	CXX0017: Error: symbol "f" not found
&d	CXX0017: Error: symbol "d" not found
&e	CXX0017: Error: symbol "e" not found
&f	CXX0017: Error: symbol "f" not found
x	CXX0017: Error: symbol "x" not found
a	2
b	3
c	6
⊞ &a	0x0018fee0
⊞ &b	0x0018fee4
⊞ &c	0x0018fee8
m	-858993460

◀▶\ **Watch1** ⟨ Watch2 ⟩ Watch3 ⟩ Watch4 ⟩

Figure 4.42: Debugging process of maximum finding program 4.

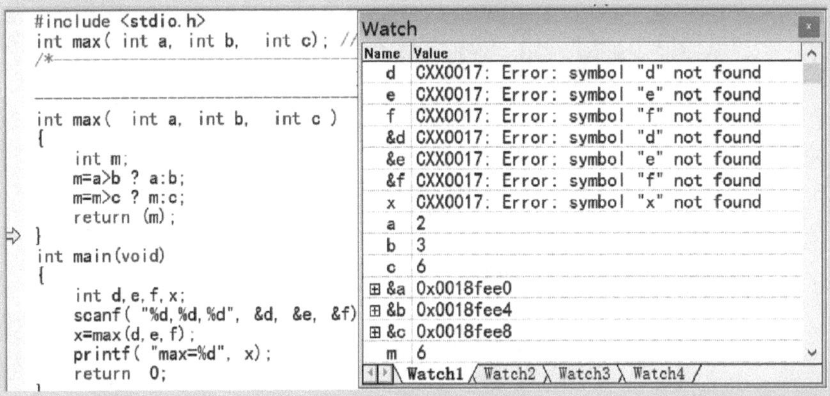

```
#include <stdio.h>
int max( int a, int b,   int c); //
/*─────────────────────────────
─────────────────────────────
int max(  int a, int b,   int c )
{
    int m;
    m=a>b ? a:b;
    m=m>c ? m:c;
    return (m);
⇨ }
int main(void)
{
    int d,e,f,x;
    scanf( "%d,%d,%d", &d, &e, &f)
    x=max(d,e,f);
    printf( "max=%d", x);
    return 0;
```

Watch	
Name	**Value**
d	CXX0017: Error: symbol "d" not found
e	CXX0017: Error: symbol "e" not found
f	CXX0017: Error: symbol "f" not found
&d	CXX0017: Error: symbol "d" not found
&e	CXX0017: Error: symbol "e" not found
&f	CXX0017: Error: symbol "f" not found
x	CXX0017: Error: symbol "x" not found
a	2
b	3
c	6
⊞ &a	0x0018fee0
⊞ &b	0x0018fee4
⊞ &c	0x0018fee8
m	6

◀▶\ **Watch1** ⟨ Watch2 ⟩ Watch3 ⟩ Watch4 ⟩

Figure 4.43: Debugging process of maximum finding program 5.

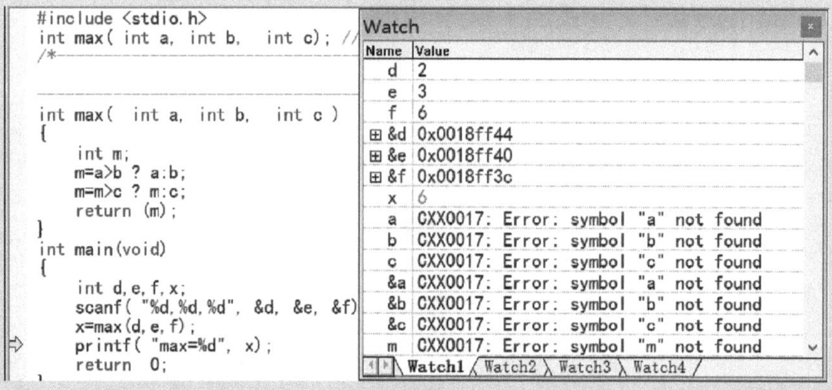

```
#include <stdio.h>
int max( int a, int b,  int c); //
/*

int max( int a, int b,  int c )
{
    int m;
    m=a>b ? a:b;
    m=m>c ? m:c;
    return (m);
}
int main(void)
{
    int d, e, f, x;
    scanf( "%d,%d,%d", &d, &e, &f)
    x=max (d, e, f);
    printf( "max=%d", x);
    return  0;
}
```

Watch		
Name	Value	
d	2	
e	3	
f	6	
⊞ &d	0x0018ff44	
⊞ &e	0x0018ff40	
⊞ &f	0x0018ff3c	
x	6	
a	CXX0017: Error: symbol "a" not found	
b	CXX0017: Error: symbol "b" not found	
c	CXX0017: Error: symbol "c" not found	
&a	CXX0017: Error: symbol "a" not found	
&b	CXX0017: Error: symbol "b" not found	
&c	CXX0017: Error: symbol "c" not found	
m	CXX0017: Error: symbol "m" not found	

◄►\ **Watch1** / Watch2 \ Watch3 \ Watch4 /

Figure 4.44: Debugging process of maximum finding program 6.

Example 4.2 Structure variable as parameter

1. Problem description

Use the debugger to analyze the characteristics of passing a structure variable when using it as the parameter.

2. Code implementation

```c
#include <stdio.h>
struct student
{ int  num;
    float grade;
};
struct student func1(struct student stu) //Structure variable as parameter
{
  stu.num=101;
  stu.grade=86;
  return (stu); //Return a structure variable
}
int main(void)
{
  struct student x={0, 0};
  struct student y;
  y = func1(x); //Structure variable as argument
  return 0;
}
```

3. Debugging

In Figure 4.45, note that address of structure argument x is 0x12ff78.

In Figure 4.46, note that the address of parameter stu is 0x12ff14, which is different for the address of x. In conclusion, the value of the argument is copied into the parameter.

```
int main()
{
    struct student x={0, 0};
    struct student y;

⇨   y = func1(x);
    return 0;
}
```

▼atch	⊠
Name	Value
⊟ &x	0x0012ff78
├ num	0
└ grade	0.000000

Figure 4.45: Structure variable as parameter debugging step 1.

```
struct student func1(struct student stu)
⇨ {
    stu.num=101;
    stu.grade=86;
    return (stu);
}
```

▼atch	⊠
Name	Value
&x	CXX0017: Err
⊟ &stu	0x0012ff14
├ num	0
└ grade	0.000000

Figure 4.46: Structure variable as parameter debugging step 2.

In Figure 4.47, members of structure stu are modified in child function func1.

In Figure 4.48, structure y in the main function is used to store the value of the structure variable returned by func1.

```
struct student func1(struct student stu)
{
    stu.num=101;
    stu.grade=86;
⇨   return (stu);
}
```

▼atch	⊠
Name	Value
&x	CXX0017: Err
⊟ &stu	0x0012ff14
├ num	101
└ grade	86.0000

Figure 4.47: Structure variable as parameter debugging step 3.

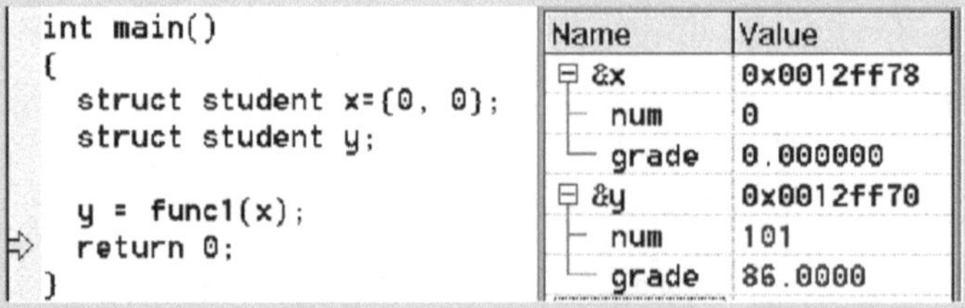

```
int main()
{
    struct student x={0, 0};
    struct student y;

    y = func1(x);
    return 0;
}
```

Name	Value
⊟ &x	0x0012ff78
─ num	0
─ grade	0.000000
⊟ &y	0x0012ff70
─ num	101
─ grade	86.0000

Figure 4.48: Structure variable as parameter debugging step 4.

Note: x, y, and stu have different addresses. The value of x is not modified.

Conclusion About call by value
(1) Parameters and arguments are stored separately;
(2) During function call, values of arguments are copied into parameters;
(3) Computation in the child function uses parameters. Updates of parameter values do not affect arguments.

In essence, call by value copies values of arguments into parameters. Thus, updates of parameters in child functions would not affect the variables used in the function call. Hence, call by value protects our data by preventing the function being called from modifying variables in the caller.

4.5.2 Call by reference

Example 4.3 Computing partial sum of array
Compute the sum of elements between indices m and n in integer array score. See Figure 4.49 for the schematic.
    The main function should read values for m and n and output results. The sum computation should be done by child function func.

Analysis
In this problem, parameters can be passed in multiple ways. We shall implement the program using three ways of parameter passing.

Index	0	1	2	3	4	5	6	7	8	9
score[ ]	1	2	3	4	5	6	7	8	9	0

(with m pointing at index 3 and n pointing at index 7)

Figure 4.49: Computing partial sum of an array.

Solution 1

1. Function structure design

Figure 4.50 analyzes the number of inputs and outputs of the child function.

	Content	Quantity	Parameter passing method	Parameter passing implementation	
Input	Information of array score	Multiple	Pass by reference	Parameter	int score[]
	Values of m and n	Single	Pass by value		int m,int n
Output	Sum of array elements between index m and index n	Single	return	Return	int type

If the parameter is a 1-d array, we can omit the array length

Figure 4.50: Key elements analysis of solution 1.

The input needs to contain all information of array score and values of indices m and n. There are multiple elements in score. m and n are both single variables. Hence, we shall pass the array score by address and pass m and n by value.

The output is the partial sum of elements between indices m and n. Because it is a single value, we can return it using a return statement. The return type is int.

2. Function implementation design

Based on the key elements, we can write out the function header. As for the function body, we can use a for loop to add elements between indices m and n into variable sum, and return the sum using a return statement, as shown in Figure 4.51.

Function header	Function type	Function name	Parameter list
	int	func	(int *sPtr, int m, int n)
Function body	{	int i, sum=0;	
		sPtr = &sPtr[m];	
		for (i= m; i<=n; i++, sPtr++)	
		sum = sum + *sPtr; //Compute sum of elements between index m and index n	
		return sum;	
	}		

Figure 4.51: Function design of solution 1.

3. Code implementation

The code implementation is given in Figure 4.52.

The child function is between line 5 and line 15, while the remaining part is the main function.

```
01 #include "stdio.h"
02 #define SIZE 10
03 int func( int score[ ], int m, int n);
04
05 //Compute the sum of elements of array score between index m and index n
06 int func(int score[ ],int m,int n)
07 {
08    int i,sum=0;
09
10    for (i= m; i<=n; i++)
11    {
12       sum=sum+score[i];
13    }
14    return sum;
15 }
16 int main(void)
17 {
18    int x;
19    int a[SIZE]= {1,2,3,4,5,6,7,8,9,0};
20    int p=3 , q=7;           //Specify range of sum
21
22    printf( "Elements of array a between index %d and index %d are:",p,q);
23    for ( int i= p; i<=q; i++)
24    {
25       printf( "%d",a[i]);
26    }
27    printf( "\n");
28    x=func(a,p,q);
29    printf( "Sum of elements of array a between index %d and index %d are: %d\n",p,q,x);
30    return 0;
31 }
```

Display specified elements

If an argument is an array, we only need to write the array name

Program result:
Elements of array a between index 3 and index 7 are: 4 5 6 7 8
Sum of elements of array a between index 3 and index 7 are: 30

Figure 4.52: Code implementation of solution 1.

Lines 22–27 print values of elements in the specified range so that it is easier to debug later.

On line 28, function func is called with argument array a and indices p and q. Pay attention to how we use the array name in the argument list.

4. Debugging

Before debugging, we should list issues we would like to investigate, which include questions related to address passing and variables we want to inspect. Figure 4.53 shows the values of these variables in the Watch and Memory windows of the debugger. During debugging, we can use a table to record the values of variables for further analysis. With all the information we have, we can conclude that call by reference uses the same memory space for parameters and arguments, while call by value uses separate memory spaces for them.

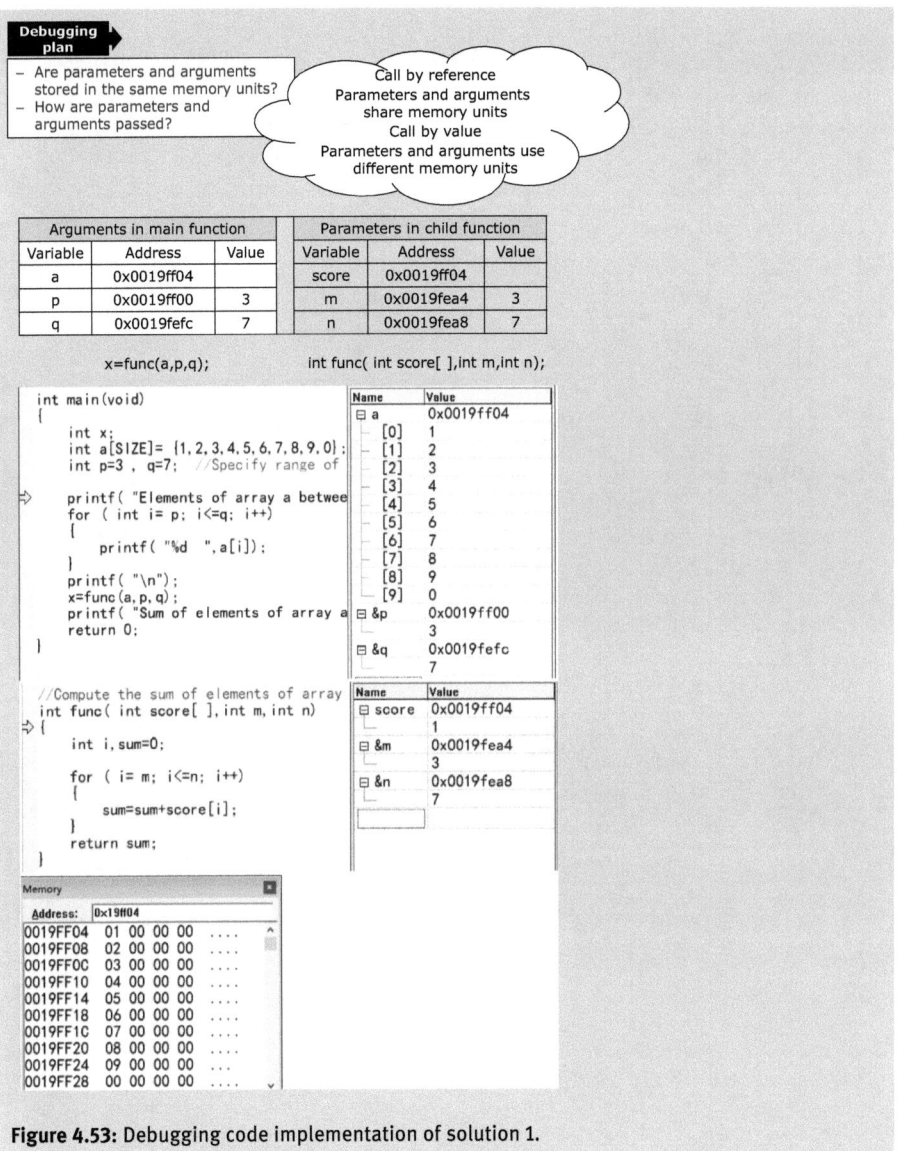

Figure 4.53: Debugging code implementation of solution 1.

Solution 2

1. Function structure design

The computation result of the child function can also be accessed using a shared address, as shown in Figure 4.54. In this case, the function type is void, and the result is stored at a specified position of array score. An integer variable size represents the position.

	Content	Quantity	Parameter passing method	Parameter passing implementation	
Input	Information of array score	Multiple	Pass by reference	Parameter	int score[]
	Values of m and n	Single	Pass by value		int m,int n
	Position in score that is used to store result	Single	Pass by value		int size
Output	Sum of array elements between index m and index n	Single	Pass by reference	Function type	void

Figure 4.54: Key elements analysis of solution 2.

2. Function implementation design

After designing the function structure of solution 2, we can write out the function header. As shown in Figure 4.55, the function body uses a for loop to compute the sum and stores sum at a specified position of array score.

Function header	Function type	Function name	Parameter list
	void	func	(int score[], int m, int n, int size)
Function body	{	int i, sum=0;	
	for (i=m; i<=n; i++)		
	sum=sum+score[i];		
	score[size]=sum; //The sum is stored in specified position of score		
	}		

Figure 4.55: Function design of solution 2.

3. Code implementation

Figure 4.56 shows the code implementation of solution 2.

```
01 #include "stdio.h"
02 #define SIZE 10
03
04 void func( int score[ ],int m,int n,int size);
05
06 //Compute the sum of elements of array score between index m and index n,
   //store result in position with index size
07 void func( int score[ ],int m,int n,int size)
08 {
09    int i,sum=0;
10
11    for ( i= m; i<=n; i++)
12    {
13       sum=sum+score[i];
14    }
15    score[size]=sum; //The sum is stored in specified position of score
16 }
17 int main(void)
18 {
19    int a[SIZE]= {1,2,3,4,5,6,7,8,9,0};
20    int p=3, q=7;    //Specify range of sum
21
22    printf("Elements of array a between index %d and index %d are:",p,q);
23    for (int i= p; i<=q; i++)
24    {
25       printf("%d ",a[i]);          Nonvalue-returning
26    }                                   functions
27    printf("\n");
28    func(a,p,q,SIZE-1);
29    printf("Sum of elements of array a between index %d and index %d are:
30    %d\n",p,q,a[SIZE-1]);
30    return 0;
31 }
```

Figure 4.56: Code implementation of solution 2.

The child function is declared and defined between line 4 and line 16. Note that the type of function is void, as shown in line 28, so it is a nonvalue-returning function call.

The main function is the same as in solution 1, except that the function call is slightly different.

4. Debugging

As usual, we need to list issues for investigation before debugging. We shall focus on issues related to passing by reference in this solution, as shown in Figure 4.57. The value of the last element of array a in the main function should be 0 before calling child function func and 30 after the call. Figure 4.58 shows debugging information of the program.

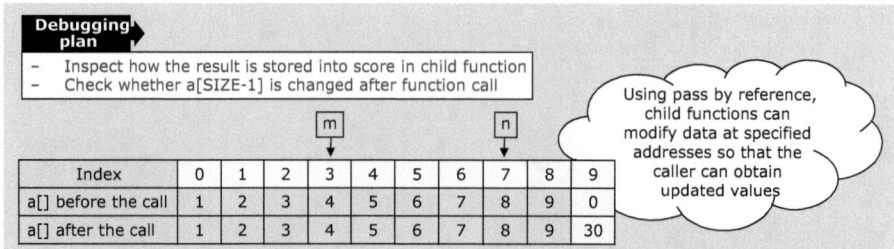

Figure 4.57: Debugging plan of solution 2.

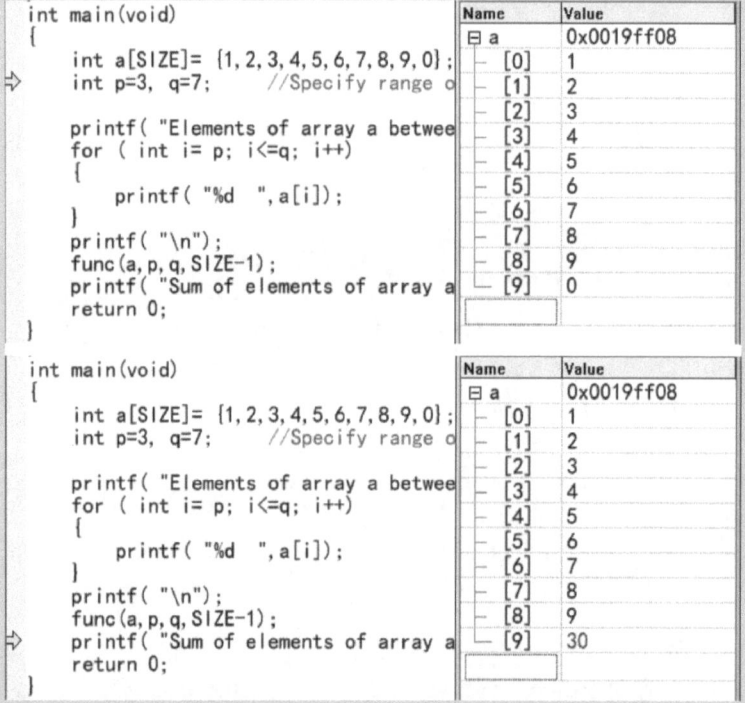

Figure 4.58: Debugging code implementation of solution 2.

Solution 3

1. Function structure design

Figure 4.59 shows the third solution. Compared with solution 1, the only difference is that the beginning address of the score is passed to the child function using a pointer. This is an alternative form of "call by reference:" using a pointer as an argument.

2. Function implementation design

Figure 4.60 shows the implementation of solution 3. The parameter uses pointer sPtr to store the address of array score. On line 2 of the function body, we point sPtr to address of the element we want to access, that is, address of the element with index m. Then we use the for loop to compute the sum. Note that we refer elements of score using pointer sPtr. Finally, the sum is returned using the return statement.

An alternative way of pass by reference: using a pointer as parameter

	Content	Quantity	Parameter passing method	Parameter passing implementation	
Input	Information of array score	Multiple	Pass by reference	Parameter	int *sPtr
	Values of m and n	Single	Pass by value		int m,int n
Output	Sum of array elements between index m and index n	Single	Pass by value	Return	int type

Figure 4.59: Key element analysis of solution 3.

Function header	Function type	Function name	Parameter list
	int	func	(int *sPtr, int m, int n)
Function body	{ int i, sum=0;		
	sPtr = &sPtr[m]; //sPtr points to address of the element to be accessed		
	for (i= m; i<=n; i++, sPtr++)		
	sum = sum + *sPtr; //Compute sum of elements of score between index m and index n		
	return sum;		
	}		

Figure 4.60: Function implementation design of solution 3.

3. Code implementation

```
01  #include "stdio.h"
02  #define SIZE 10
03  int func( int *sPtr,int m,int n);
04
05  int func( int *sPtr, int m, int n)
06  {
07    int i,sum =0;
08
09    sPtr = &sPtr[m]; //Point sPtr to address of element to be accessed
10    for ( i= m; i<=n; i++, sPtr++)
11    {
12      sum = sum + *sPtr;
13    }
```

```
14    return sum;
15 }
16 int main(void)
17 {
18    int x;
19    int a[SIZE] = {1,2,3,4,5,6,7,8,9,0};
20    int *aPtr = a;
21    int p=3,  q=7; //Specify the range
22
23    x=func(aPtr,p,q);
24    printf( "%d\n",x);
25    return 0;
26 }
```

4. Debugging

As usual, we list issues related to pass by reference and variables we want to inspect, as shown in Figure 4.61. Argument aPtr and parameter sPtr should have the same value, which ought to be the address of array a.

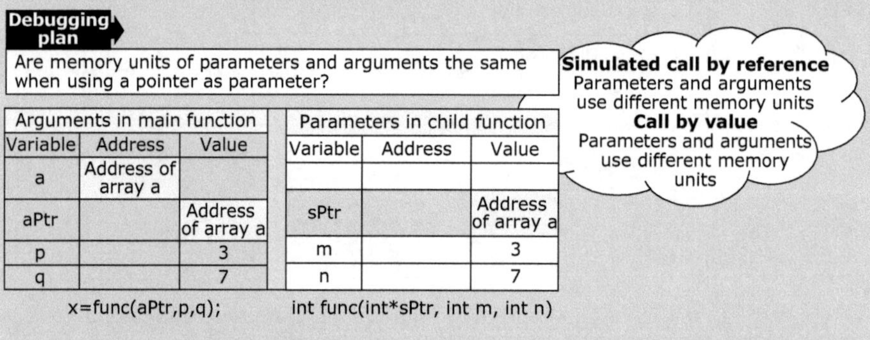

Debugging plan

Are memory units of parameters and arguments the same when using a pointer as parameter?		

Arguments in main function			Parameters in child function		
Variable	Address	Value	Variable	Address	Value
a	Address of array a				
aPtr		Address of array a	sPtr		Address of array a
p		3	m		3
q		7	n		7

x=func(aPtr,p,q);   int func(int*sPtr, int m, int n)

Simulated call by reference
Parameters and arguments use different memory units
Call by value
Parameters and arguments use different memory units

Figure 4.61: Debugging plan of solution 3.

Figure 4.62 shows the first step of debugging.

In the main function, the address of array a is 0x18ff1c. aPtr is a pointer pointing to the address of a, so their values should be identical. The address of aPtr can be obtained using & operator. We can see in the Watch window that the address is 0x18ff18.

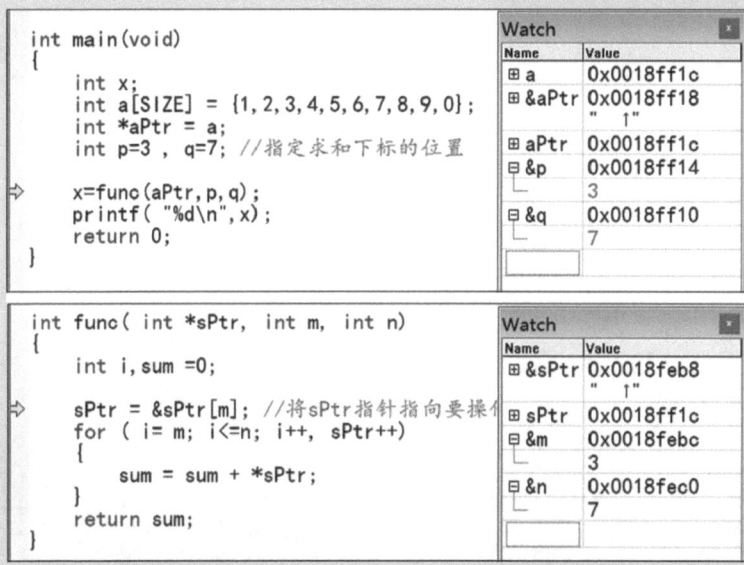

Figure 4.62: Debugging solution 3 step 1.

In the child function func, the address of sPtr is 0x18feb8 and the value of sPtr is the same as array a.

With the information displayed in the Watch window, we can complete the table shown in Figure 4.63. In conclusion, parameter addresses and argument addresses are different when passing a pointer to the child function. This is similar to call by value. We call such a function call a simulated call by reference.

Arguments in main function			Parameters in child function		
Variable	Address	Value	Variable	Address	Value
a	0x0018ff1c		sPtr	0x0018feb8	0x0018ff1c
aPtr	0x0018ff18	0x0018ff1c	m	0x0018febc	3
p	0x0018ff10	3	n	0x0018fec0	7
q	0x0018ff14	7			
x=func(aPtr,p,q)			int func(int*sPtr, int m, int n)		

Figure 4.63: Debugging information table of solution 3.

We continue the debugging process until sPtr is going to point to the element with index 3, as shown in Figure 4.64. Note that the value of sPtr is 0x18ff1c at this moment. Then we enter the for loop with i = 3. The value of sPtr is updated to 0x18ff28. Using the asterisk operator, we can obtain the value of the unit it points to, which is 4. We can also inspect the memory layout at this address in the Memory window.

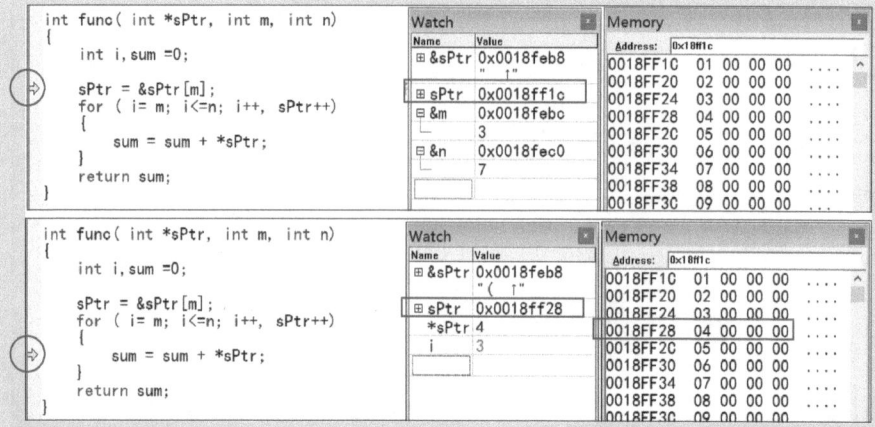

Figure 4.64: Debugging solution 3 step 2.

In the next iteration, as shown in Figure 4.65, we have i = 4 and sPtr pointing to 0x18ff2c after increasing by 1. The value stored at this address is 5. In the next iteration, i becomes 5 and sPtr points to address 0x18ff30, in which value 6 is stored.

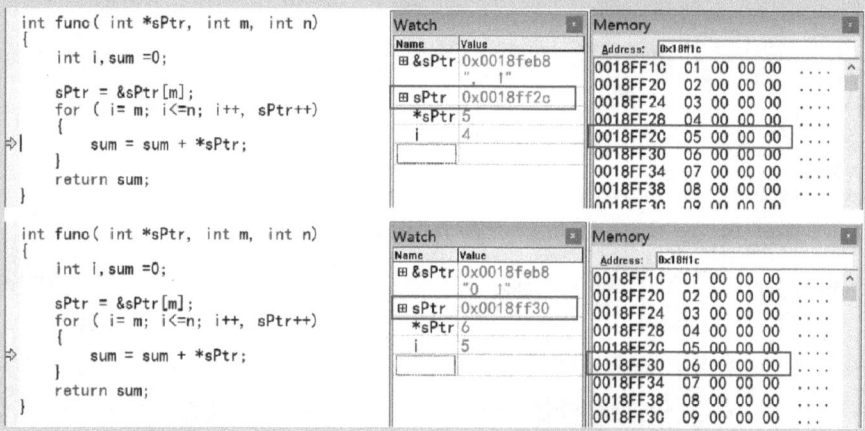

Figure 4.65: Debugging solution 3 step 1.

When the value of i becomes 8, the loop terminates, as shown in Figure 4.66. Note that sPtr now points to address 0x18ff3c. The function func then terminates, and the program returns to the main function. The computation result 30 is now stored into x. It is worth noting that the address and value of aPtr did not change as sPtr changed.

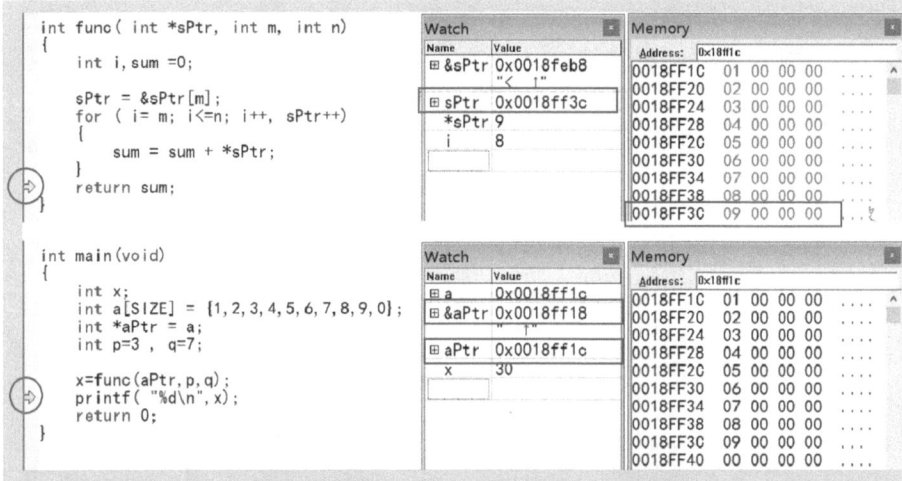

```
int func( int *sPtr, int m, int n)
{
    int i, sum =0;

    sPtr = &sPtr[m];
    for ( i= m; i<=n; i++, sPtr++)
    {
        sum = sum + *sPtr;
    }
    return sum;
}

int main(void)
{
    int x;
    int a[SIZE] = {1, 2, 3, 4, 5, 6, 7, 8, 9, 0};
    int *aPtr = a;
    int p=3 , q=7;

    x=func(aPtr, p, q);
    printf( "%d\n", x);
    return 0;
}
```

Figure 4.66: Debugging solution 3 step 4.

Conclusion About call by reference and call by value

Figure 4.67 summarizes the three types of calls and their characteristics. In C programs, arguments and parameters are stored in different memory units when the variables passed are numbers or pointers. They share the same space only when an array is passed.

The merit of using call by reference is that the information transmission efficiency becomes higher because fewer data are copied during the process.

Conclusion

	Type	Memory units of parameters and arguments	Information transmission direction	Call type
Parameter	Value	Different	Single	Call by value
	Pointer	Different	Double	
	Array name	Shared	Double	Call by reference

Figure 4.67: Summarization of variable passing rules.

4.5.3 Comprehensive examples of functions

Example 4.4 Calling the same function multiple times

Please write a program that computes the number of k-combinations from n elements.

Analysis

As shown in Figure 4.68, the formula requires the computation of multiple factorials. Consequently, we can write a function that computes factorial and reuse it.

$$nC_k = \frac{n!}{k! \times (n-k)!} \qquad (n > k)$$

Reuse factorial function

Figure 4.68: Formula of computing k-combinations.

1. Function structure design

Key elements of the function are shown in Figure 4.69. The function name is factorial. Given an integer x, the function outputs factorial of x. If the input is invalid, the function should return -1.

Functionality	Input information	Output information	
Compute factorial	int x	int value	Exception: −1
			Normal: >0
Function name	Parameter list	Function type	

Figure 4.69: Key elements of function factorial.

We can write out the function header based on these key elements, and the function body based on its functionality. An exception handling routine is necessary so that the function returns −1 when x<0. The cumulative product computed in the for loop is stored in variable f.

Although there are two return statements, the function has to terminate through one and exactly one exit, as shown in the flowchart (Figure 4.70).

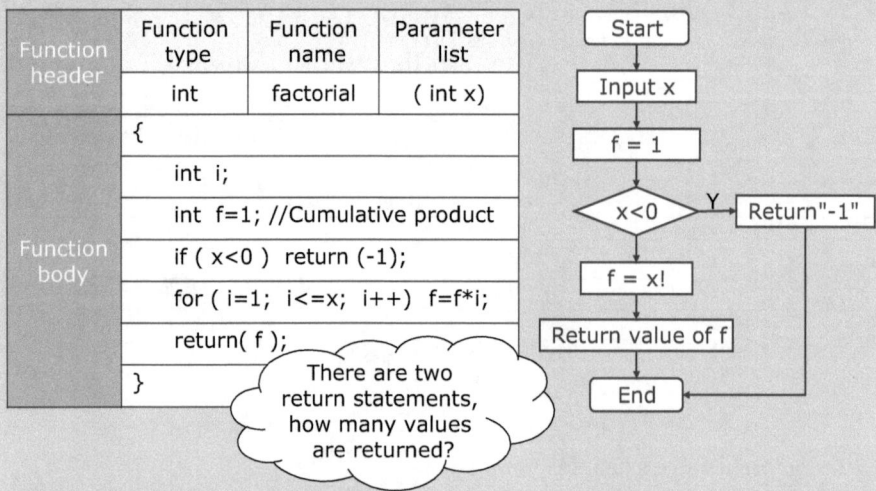

Figure 4.70: Structure and function body design of factorial.

2. Code implementation

As shown in Figure 4.71, the main function calls factorial multiple times in one expression.

```
#include <stdio.h>
int factorial (int x);                                     Function declaration

int factorial (int x)
{
    int  i;
    float t=1;                                             Function definition
    for(i=1;  i<=x;  i++) t=t*i;
    return (t);
}

int main(void )
{
    int c;
    int m,n;

    printf("input m,n:");
    scanf(" %d%d",&m, &n);
    c=factorial (m)/(factorial (n)*factorial (m-n));       Function call
    printf("The result  is %8.1f", c);
    return 0;
}
```

Figure 4.71: Multiple calls of the factorial function.

Example 4.5 Calling multiple functions
There is a sorted array. Please write a program that inserts an input number into the array so that the result array is still sorted.
    Requirement: use binary search to find the insertion position and then use a move function to move elements backwards. The sorted array and the input is given in the main function.

1. Algorithm design
Regardless of finding the keyword or not using a binary search function, the mid value will be the index of the position at which the new number will be inserted. Having found this index, we can use the move function to move array elements backwards and insert the number at position mid.

2. Code implementation
```
/*================================================
Functionality: binary search
Input: address of sorted array, array length, value of keyword to be found
Output: position of the last search
=============================================*/
int BinarySearch (int a[], int n, int key)
{
  int low=0, high=n-1;
  int mid;
  while (low<=high)
  {
    mid = (low+high+1)/2;
    if (a[mid]== key) break; //Search succeeded
    else
```

```
  {
    if (a[mid]> key) high = mid-1; //Search in the low range
    else low = mid+1; //Search in the high range
  }
 }
 return mid;
}
/*================================================
Functionality: move array elements
Input: address of array, array length, position from which the move starts
Output: None
=================================================*/
void move(int a[], int n, int subscript)
{
 int i;
 for(i=n;i>subscript;i--)
 {
  a[i]=a[i-1];
 }
}

#include <stdio.h>
#define N 11
int main(void)
{
 int array[N]={5,10,19,21,31,37,42,48,50,55};
 int number;        //Number to be inserted
 int insert_sub; //Insert position
 printf("The original array:\n");
 for(i=0;i<N;i++) printf("%d ",array[i]);
 printf("\n");
 printf("Please insert a new number:");
 scanf("%d",&number);
 insert_sub=BinarySearch (array,N-1,number); //Compute insert position
 move(array,N-1,insert_sub+1); //Move elements after insert position afterwards
 array[insert_sub+1]=number; //Insert the number
 printf("The array after insertion:\n");
 for(i=0;i<N;i++) printf("%d ",array[i]);
 printf("\n");
 return 0;
}
```

Example 4.6 Nested function calls
Compute the difference of the maximum and the minimum of three numbers.

Analysis
Although the problem is trivial, we shall use three functions to solve it in order to demonstrate nested function calls. The code implementation is as follows:

```
int dif(int x,int y,int z);   //Compute the difference of the maximum and
                              //the minimum of x, y and z
int max(int x,int y,int z);   //Compute the maximum of x, y and z
int min(int x,int y,int z);   //Compute the minimum of x, y and z
int main(void)
{
  int a,b,c,d;
  scanf("%d%d%d",&a,&b,&c);
  d=dif(a,b,c);
  printf("Max-Min=%d\n",d);
  return 0;
}
int dif(int x,int y,int z) //Compute the difference of the maximum and
                           //the minimum of x, y and z
{
  return (max(x,y,z) - min(x,y,z));
}
int max(int x,int y,int z) //Compute the maximum of x, y and z
{
  int r;
  r= x>y ? x:y;
  return(r>z?r:z);
}
int min(int x,int y,int z) //Compute the minimum of x, y and z
{
  int r;
  r = x<y ? x:y;
  return(r<z ? r:z);
}
```

Example 4.7 Two-dimensional array as parameter
Find the highest grade from grades of three students in four courses.

Analysis
1. Data structure design
We shall use a two-dimensional array studentGrades[number of students][number of courses] to store all the grades.

2. Function design
Based on the problem description, we can summarize the key elements of the function, as shown in Figure 4.72.

Function name	Functionality	Parameter	Function type
maximum	Determine the highest score	Grade table, the number of students, the number of courses	int

Figure 4.72: Key elements of student grades processing function.

3. Code implementation

```c
//Process 2-dimensional array in child function
#include <stdio.h>
#define STUDENTS 3
#define EXAMS 4
//Function declaration, see section 4.6.6 for introduction of const
int maximum( const int grades[ ][EXAMS], int pupils, int tests );
//When using 2-d array as parameter, the row size can be omitted in
//definition and declaration, but the column size cannot
int main(void)
{
  //Initialize students' grades
  int studentGrades[STUDENTS][EXAMS]
  = { { 77, 68, 86, 73 },
  { 96, 87, 89, 78 },
  { 70, 90, 86, 81 }
  };
  printf( "Highest grade: %d\n",maximum(studentGrades,STUDENTS,EXAMS));
  return 0;
}
int maximum(const int grades[ ][EXAMS], int pupils, int tests )
{
  int i;   //Counter of student
  int j;   //Counter of courses
  int highGrade = 0; //Initialize with lowest possible grade
```

```
for ( i = 0; i < pupils; i++ ) //Iterate through rows
{
  for ( j = 0; j < tests; j++ ) //Iterate through columns
  {
    if ( grades[i][j] > highGrade )
    {
      highGrade = grades[i][j];
    }
  }
}
return highGrade; //Return highest score
}
```

Example 4.8 Structure array as parameter
Write a function output() to print records of five students.

Analysis
1. Algorithm design
The student records are stored in a structure array student stu[], which is passed to child function output() by the main function through passing its address.

2. Code implementation
```c
#include <stdio.h>
#define N 5
struct student
{
 int num;
 char name[8];
 int score[4];
};
void output(struct student stu[])
{
 int i,j;
 printf("\nNo. Name  Sco1 Sco2 Sco3\n"); //Print table header
 for (i=0; i<N; i++)
 {
  printf("%-6d%-6s",stu[i].num,stu[i].name); //Print ID and name
  for (j=0; j<3; j++) printf("%-6d",stu[i].score[j]); //Print grades
  printf("\n");
 }
}
```

```
int main(void)
{
 struct student stu[N]=
 {
  {1001,"zhao",98,78,86,76},
  {1002,"qian",92,68,76,67},
  {1003,"sun",78,65,81,72},
  {1004,"li",91,73,85,74},
  {1005,"zhou",90,73,85,71},
};
 output(stu);
 return 0;
}
```

Example 4.9 Pointer as return value

Analysis

1. Code implementation
```
#include <stdio.h>
struct student
{ int  num;
  float grade;
};
struct student* func2(struct student stu)
{
   struct student *str=&stu;
   str->num=101;
   str->grade=86;
   return (str); //Return structure pointer
}
int main( )
{
   struct student x={0, 0};
   struct student *stuPtr;
   stuPtr = func2(x);
   return 0;
}
```

2. Debugging
The process of debugging this program is shown further. In Figure 4.73, the address of argument x is 0x12ff78.

```
int main()
{
    struct student x={0, 0};
    struct student *stuPtr;

    stuPtr = func2(x);
    return 0;
}
```

Name	Value
⊞ &x	0x0012ff78
⊞ stuPtr	0xcccccccc

Figure 4.73: Pointer as return value program debugging step 1.

In Figure 4.74, the address of the parameter stu is 0x12ff20.

```
struct student* func2(struct student stu)
{
    struct student *str=&stu;
    str->num=101;
    str->grade=86;
    return (str);
}
```

Watch ×

Name	Value
⊟ &stu	0x0012ff20
num	0
grade	0.000000

Figure 4.74: Pointer as return value program debugging step 2.

In Figure 4.75, the values of members in the structure stu are updated.

```
struct student* func2(struct st
{
    struct student *str=&stu;
    str->num=101;
    str->grade=86;
    return (str);
}
```

Name	Value
⊞ &stu	0x0012ff20
⊟ str	0x0012ff20
num	101
grade	86.0000

Figure 4.75: Pointer as return value program debugging step 3.

In Figure 4.76, stuPtr in the main function is used to store the value of local variable str. A local variable is a variable defined inside a function.

```
int main()
{
    struct student x={0, 0};
    struct student *stuPtr;

    stuPtr = func2(x);
    return 0;
}
```

Name	Value
&stu	Error: cann(
str	CXX0017: Err
⊟ stuPtr	0x0012ff20
num	101
grade	86.0000

Figure 4.76: Pointer as return value program debugging step 4.

Note: It is not recommended to return the addresses of local variables. Because the system reclaims the memory space of local variables after the function returns, information related to local variables is no longer guaranteed to be correct.

Example 4.10 Void pointers as return value
Please define a dynamic array to store grades of n students and compute the average grade. The number of students and grades are read from keyboard input.

Analysis

1. Background knowledge

The array definition method introduced in chapter "Array" allocates memory statically. In other words, the size of the array and the address it is stored at are unchanged during program execution. Suppose that we want to insert new data into the array during program execution, but the allocated array space is already fully used. Is there a way to expand the array space? There is a memory allocation method called "dynamic memory allocation" in C: if a program needs extra storage space during execution, it can "request" memory space of a certain size. When the program no longer needs the space, the space can be returned to the system. Related library functions include malloc(), calloc(), free(), and realloc(). One must include the header file stdlib.h or malloc.h to use these functions.

    1)   Memory allocation function malloc()

Prototype: void *malloc(unsigned size);
Functionality: allocates a block of memory of size bytes.
Parameters: size is an unsigned integer, which represents the size of the requested memory space.
Return value: the address of the newly allocated memory is returned. If there is no memory available, NULL shall be returned.

    Note:

(1)   NULL is returned when size is 0.
(2)   void* is a typeless pointer that can point to memory units of any type. A typeless pointer can be assigned to pointers of other types after forced type conversion.

    2)   Memory release function free()

Prototype: void free(void *block);
Functionality: releases memory space allocated using calloc(), malloc(), and realloc().
Parameter: block is a void pointer pointing to the memory to be reclaimed.
Return value: there is no return value.
Usage: void free(void *p);
The statement above releases memory in dynamic memory space pointed to by p, which is a value returned by malloc(). Free function has no return value.

2. Code implementation

```
1  #include <stdio.h>
2  #include <malloc.h>
3
4  int *DefineArray(int n);   /*Define a dynamic array of size n*/
5  void FreeArray(int *p);    /*Release memory pointed to by p*/
6
7  int main()
8  {
```

```
9    int *p, i;
10   int nCount;   /*Number of students*/
11   float fSum=0;   /*Total grade*/
12
13 /*Input number of students*/
14 printf("\nPlease input the count of students: ");
15 scanf("%d",& nCount);
16
17 /*Define a dynamic array p*/
18 p= DefineArray(nCount);
19 if (p==NULL) return 1;   /*Exception routine*/
20
21 /*Input grades of each student*/
22 printf("Please input the scores of students: ");
23 for( i=0; i< nCount; i++ )
24 {
25   scanf("%d", &p[i]);
26 }
27
28 /*Compute total grade*/
29 for(i=0;i< nCount;i++)
30 {
31   fSum+=p[i];
32 }
33
34 /*Print average grade*/
35 printf("\nAverage score of the students: %3.1f", fSum/nCount);
36
37 /*Free dynamic array p*/
38 FreeArray(p);
39 return 0;
40 }
41
42 /*Dynamically request memory space of size n*sizeof(int), which is used
for an int array with n elements*/
43 int *DefineArray(int n)
44 {
45   return (int *) malloc( n*sizeof(int) );
46 }
47
```

```
48 /*Release memory allocated by malloc*/
49 void FreeArray(int *p)
50 {
51  free(p);
52 }
```

Program result:

Please input the count of students: 5

Please input the scores of students: 87 97 77 68 98

Average score of the students: 85.4

Note: The input parameter n of function DefineArray is the number of elements in the array. The function uses malloc to allocate memory required by the array and casts the void pointer returned into integer pointer. Finally, the function returns a pointer pointing to an integer variable or array. The value of the pointer is precisely the beginning address of the memory space allocated by malloc.

i

Knowledge ABC Memory leak

Applications usually use functions like malloc or realloc to obtain blocks of memory from the heap. When the memory is no longer needed, programs must call function free to free these memory blocks; otherwise, they cannot be used again. In this case, we consider these memory blocks "leaked."

A memory leak affects the performance of a computer by reducing the amount of available memory in it. In the worst case, most of the available memory may eventually become allocated, and all or some of the devices stop working correctly, or the application fails.

In modern operating systems, standard memory used by an application is released when it terminates. In other words, a memory leak in a program with a short lifespan is rarely severe.

Much more serious leaks include those:

(1) where the program runs for an extended time and consumes additional memory over time, such as background tasks on servers, but especially in embedded devices which may be left running for many years;

(2) where new memory is allocated frequently for one-time tasks, such as when rendering the frames of a computer game or animated video;

(3) where the program can request memory (such as shared memory) that is not released, even when the program terminates;

(4) where the leak occurs within the operating system;

(5) when a system device driver causes the leak;

(6) where memory is limited, such as in an embedded system or portable device;

(7) running on an operating system (such as AmigaOS) that does not automatically release memory on program termination, where memory has to be reclaimed through system restart.

Program reading exercise

Read and analyze the program, then fill in the table.

```
#include "stdio.h"
#include "string.h"
```

```
void i_s(char in[], char out[]);
void i_s( char in[], char out[])
{
  int i, j;
  int l=strlen( in);
  for (i=j=0; i<l; i++, j++)
  {
    out[j]= in[i]; //Step 1
    out[++j] = '__'; //Step 2
    in[i] += 1; //Step 3
  }
  out[j-1]= '\0';
}
int main(void )
{
  char s[]= "1234";
  char g[20];
  i_s( s, g );
  printf("%s\n", g);
  return 0;
}
```

Having learned rules of information transmission between functions, readers can try to analyze this program on their own. Figure 4.77 lists intermediate results of each step. If readers find it hard to analyze the program by merely reading it, it is also possible to use a debugger to inspect variable values.

Index	0	1	2	3	4	5	6	7
(s[]) in[]	1	2	3	4	\0			
(g[]) out[]	1							
in[] in step 3	2							

Figure 4.77: Program reading exercise.

4.5.4 Parameters of the main function

4.5.4.1 Introduction

We mentioned before that Mr. Brown wrote an arithmetic questions program for his son Daniel. The program could generate random arithmetic problems and determine whether Daniel's answer was correct. Daniel spent much time on it and enjoyed it, so Mr. Brown wanted to recommend the program to his nephew Annie, who was living in another city. Mr. Brown knew that Annie's parents were not familiar with installing programs, so he only sent a .exe file through email. The executable file required no C compiling environment so that Annie could run the program from file explorer directly. When he tested the executable file, however, he found that the console window popped up after the program is started and then disappeared quickly before he could even see the result.

"What should I do?" Mr. Brown thought to himself. He then recalled a method used before graphical user interfaces were invented. Back in those days, DOS was the dominating operating system. In DOS, all commands were sent to computers from keyboard input, including the execution of applications. The interface used for command input was all black. It is precisely the popped-up window which displays result after we execute programs in VC6.0 IDE, namely the console. The console is also called a command line interface, in which users type in commands for applications in the command line environment. Program results are also displayed in the command line interface.

However, the graphical user interface has been the de-facto standard nowadays. Is there a way to fall back to "console?" The Windows system does preserve this function. We can enter the command line interface by typing cmd in the Run window of Windows, as shown in Figure 4.78. Even if a computer is not equipped with a compiling environment, we can still run console applications in cmd command line interface. After a console application terminates, it returns to cmd so that we can inspect its result.

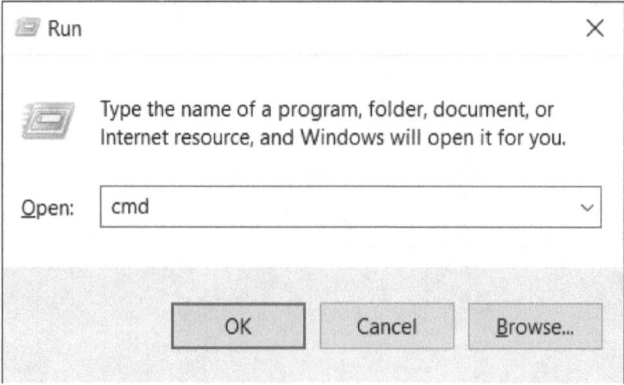

Figure 4.78: "Run" in Windows.

4.5.4.2 Parameters of the main function

We have seen the following form of the main function, which has no parameter:

```
int main( void )
{
    . . .
    return 0;
}
```

The return type of the main function is int, which is consistent with the return statement at the end of the program. 0 is the return value of the main function. Where is
it returned to then? After the main function terminates, the return value is sent
back to the operating system, indicating that the program terminates normally.

We can use parameters in plain functions, but can we do the same with the
main function? If a function has parameters, then we have to pass arguments when
calling it. However, no function can pass arguments to the main function because it
cannot be called by any function. As a result, the argument must be provided exter-
nally. How can we do this?

A C program turns into an executable file with extension .exe after being compiled and linked. An executable file can be executed directly in the operating system. In other words, it is the system that runs the file. Since other functions cannot
call or pass arguments to the main function, it has to be done by the system. In C
programs, we can pass arguments to main functions by typing them in the command line interface.

Let us take a look at the syntax of the main function with parameters:

```
int main(int argc, char *argv[])
{
    . . .
    return 0;
}
```

Command line arguments are also called positional arguments. They can be passed
to programs. Value of argc (argument count) is equal to the total number of positional arguments (including the program name). argv (argument value) is a pointer
array, in which program name is stored in argv[0] and the ith positional argument
is stored in argv[i], up until argv[argc-1]. In this way, we can pass command line
arguments into C programs without using input statements.

4.5.4.3 Example of the main function with parameters

Example 4.11 Compute rectangle area using command line inputs

Analysis
The code implementation is shown in Figure 4.79.

On line 13, the sscanf function is also an input function, which is similar to scanf. We have learned that scanf uses keyboard input (stdin). sscanf, on the other hand, uses fixed strings as inputs. Readers can refer to Appendix C of Volume 1 for more on input functions. sscanf reads data in a specified format from a string. Here it reads data from argv[1] and puts it into variable w.

```
01 #include <stdio.h>
02 #include <stdlib.h>
03
04 int main( int argc, char*argv[] )
//arg c is the number of parameters; arg v[0] is the program name, other parameters are stored after it
05 {
06    float w,h;      // Width and height of rectangle
07    if(argc< 3)   // Parameters less than 3
08    {
09        printf("input:File_Name width height\n");
10        printf("E.g.: %s 3.2 4.5\n",argv[0]);
11        exit(0);      //Exit the  program
12    }
13    sscanf(argv[1],"%f",&w); //Width
14    sscanf(argv[2],"%f",&h); //Height
15    printf("area = %f\n",w*h);
16    return 0;
17 }
```

Exception routine

There are two parameters besides program name

Figure 4.79: Example of the main function with parameters.

On line 11, exit() is a library function whose header file is either stdlib.h or windows.h. It closes all files and terminates the current process. In the statement exit(x), the value of x represents exit status and is returned to the operating system. If x is 0, the program exited normally; otherwise, it exited with exceptions.

Figure 4.80 shows the command line interface opened by cmd command. We first enter the directory of the executable file with command cd, which stands for "change directory." In this example, the directory is "D:\MyWin32App\Win32App\Debug" and the executable file is "demo.exe." Then we type in command line arguments of the main function in the interface. After typing in the program name, width, and height (the program specifies the order in which arguments are input), the program outputs area of the corresponding rectangle. We test the program with three groups of inputs, two of which are valid. In the case of invalid input, the program outputs the format of valid input and an example in the exception routine.

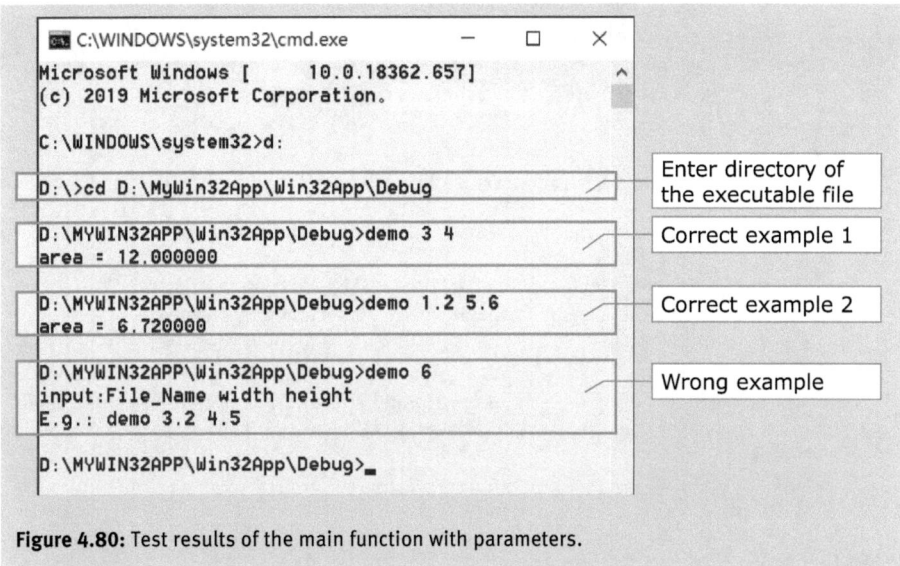

Figure 4.80: Test results of the main function with parameters.

Knowledge ABC Differences between exit() and return

According to ANSI C, they are equivalent in the first call of main().

Note that we have used the term "first call." If main() is in a recursive program, then exit() still terminates the program; on the other hand, a return statement returns to the previous level of recursion. Only at the top level of recursion does a return statement terminate the program. Besides, exit() terminates a program even if it is used in a function other than main().

Knowledge ABC Format of the main function

In the latest C99 standard, only the following two definitions of main( ) are correct (see ISO/IEC 9899:1999 (E) 5.1.2.2.1 program startup):

```
int main( void ) // without parameter
{
  . . .
  return 0;
}
int main( int argc, char *argv[] ) // with parameter
{
  . . .
  return 0;
}
```

int indicates the return type of main( ). Information passed to functions is normally written inside parentheses after function names. void means that no arguments should be passed to main( ). However, we often find the following forms of main( ) in legacy C code:

(1) main( )

This is allowed in C90 standard, but not in C99. Hence, do not write this even if it is valid in your compiler.

(2) void main( )

This is valid in some compilers, but no standard considers accepting it. Bjarne Stroustrup, the creator of C + +, makes a clear statement in the FAQ section on his website: void main( ) has never existed in C++ or C. Consequently, compilers may reject such code. In fact, it is invalid in many compilers.

The merit of sticking with the standard is: a program can normally run even after being ported from one compiler to another. In other words, it results in "better portability."

4.6 Scope

When solving practical problems with programs, the scale of programs becomes larger as problems become more complicated. This leads to many issues in programming. In response to these issues, we introduced the idea of modularization. To be more specific, we introduced functions into the C language. Figure 4.81 shows some issues related to functions, which we have seen in previous sections.

Issues we need to consider when solving problems with functions	Information transmission between functions	What should we do when the scale of a program is large?
	Definition of functions	
	Calling methods of functions	

Figure 4.81: Issues related to function design.

4.6.1 Introduction

4.6.1.1 Cooperation issues in teamwork

Mr. Brown's research group received funding for a new project. For higher efficiency, the project was divided into multiple modules, each of which was assigned to a teacher or a student in the group.

However, they realized that there were issues of cooperation that needed to be settled before starting coding.

A student said, "In small programs I have written before, child functions and main function are in the same file. However, that would be inconvenient in the case of teamwork! I think we should use a separate file for each module." Another student said, "I prefer using i, j, and k for loop control variables. Can I still use them if Prof. Brown uses them as well? Should we discuss with each other before defining our variables?"

A teacher thought for a while and responded, "From the perspective of the overall workflow, we should create program files on our own, but a program cannot run if it has no main function. From the perspective of program execution, however, we are writing one large program, which should contain only one main function. How should we do this?"

Mr. Brown summarized everyone's questions, as shown in Figure 4.82. Then he said, "To answer these questions, we need to introduce new rules in the programming language. Can you imagine what the best mechanism of working in team is? I think we should work in different files. Variables in different files can have the same name. One program should have only one main function." Everyone nodded.

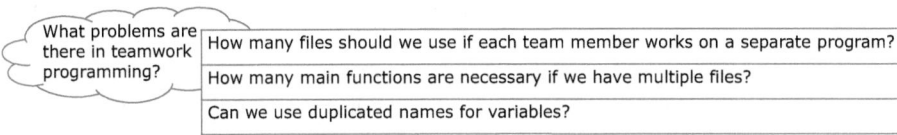

Figure 4.82: Issues in large-scale programs.

He asked further, "In this case, what mechanisms of program execution are necessary to make what I just said possible?" "I think we can attach scope and lifespan to variables so that they are isolated in a file or a function. This can stop them from messing around," another teacher answered. Everyone agreed with him.

As they have imagined, the rule in C is: a C program can consist of multiple files, each of which can have multiple functions, as shown in Figure 4.83. Variables in different files can have the same name. A C program must have one and only one main function.

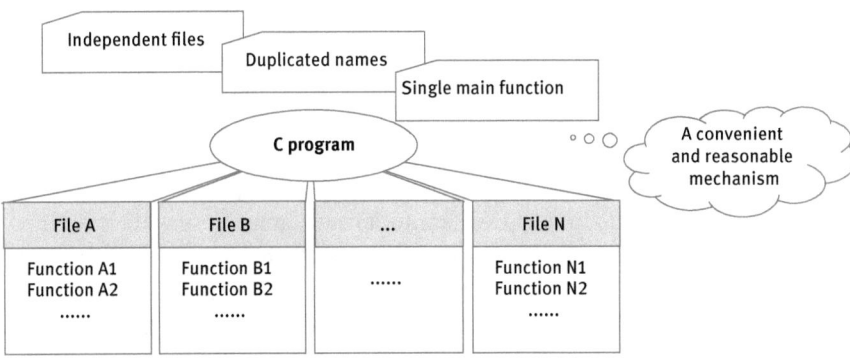

Figure 4.83: Structure of C programs.

4.6.1.2 Outsourced projects in a flow

In the rostrum building flow, some work was outsourced. Each service provider can be considered as a child function that completes a certain task. Information related to one service provider, such as material, size, or price, need not and should not be exposed to other service providers. This is the scope of information. Such a scoping strategy is also used in modularization of programs.

4.6.1.3 Resource-sharing problem

Mr. Brown's university has facilities like assembly halls, libraries, canteens, and infirmaries. There are also service departments that provide specific services. As per government's and university's policies, assembly halls and libraries are open to the public. Organizations and individuals can use them following some processes. On the other hand, canteens and infirmaries are only available to students and university employees. Depending on availability, resources can be divided into internal resources and shared resources.

The code is also a resource. If a program consists of multiple source files, program designers should be able to determine whether a function can be called by functions in other files based on the nature of the problem to be solved. In other words, the "availability" attribute should be necessary for functions as well. Depending on availability, functions in C can be divided into internal functions and external functions.

To borrow books from university libraries, individuals should present their university ID cards. These cards are also required in other facilities, such as canteens and infirmaries. In many cases, however, data regarding IDs used in one facility are only available to the management department of that facility. Departments seldom share their data. To sum up, some data are available to all departments and some are restricted to certain departments.

Imagine a department as a function. Then there are two types of information processed by functions. Data that are available to all functions are called global data, while those available only inside a function are called local data.

4.6.2 Masking mechanism of modules

Recall the idea of modularization: we aim to hide internal implementation and data of modules from outside and to ensure that modules communicate with other modules through information interfaces. To design such a mechanism, where should we start? Based on the discussion in the introduction part, it is clear that we should start from the isolation of internal data and the masking mechanism of functions.

4.6.2.1 Isolation of internal data

By masking data in a child function, which are mostly variables, we prevent them from being accessed by other functions. Issues we need to consider about these variables are shown in Figure 4.84.

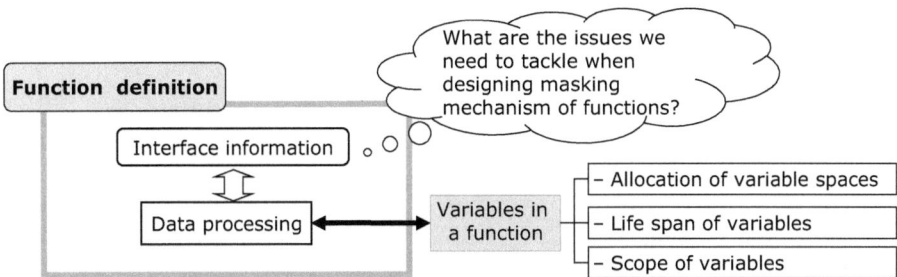

Figure 4.84: Issues related to masking internal information of functions.

4.6.2.2 Masking rule of functions

Depending on "availability," functions in C are divided into two types: internal functions and external functions. They are identified with special signs, as shown in Figure 4.85. We shall introduce these signs in the section "scope of functions."

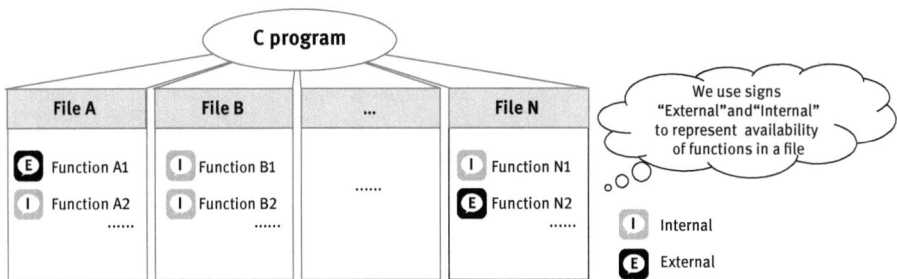

Figure 4.85: Sharing mechanism of functions in a multiple-file structure.

4.6.3 Memory segments and storage classes

4.6.3.1 Memory segments of programs

In practice, we isolate data through categorizing data and manage them differently based on the availability of data. We use the same strategy in computers to manage data and code. Figure 4.85 shows the memory layout of a C program. The code segment contains binary code of functions; the constant segment contains string literals and other constants; the dynamic segment is used to store internal data of functions, namely local variables. The static segment stores data that are shared among functions, that is, global variables.

4.6.3.2 Storage classes of variables

To distinguish variables stored in different segments, C attaches another attribute to them, that is, storage class. It also indicates the lifespan of variables in memory (Figure 4.86) and scope of variables in programs, as shown in Figure 4.87.

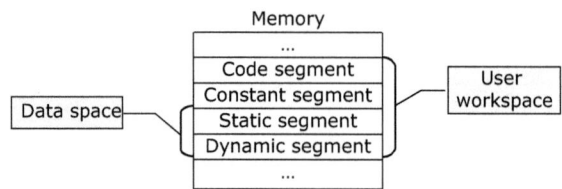

Segment		Content	Notes
Dynamic segment	Stack	Local variables, parameters	Allocated and released by the system automatically
	Heap	Memory requested using dynamic allocation functions	– Allocated and released by programmers – The system can reclaim the memory if programmers didn't release it
Static segment		Global variables, static variables	Allocated and released by the system automatically
Constant segment		Constants	Allocated and released by the system automatically
Code segment		Program code	

Figure 4.86: Memory layout.

Storage segment	Life span	Storage class	Type	Notes
Dynamic	Same as function	register	Register	Variables stored in registers
		auto	Auto	Auto variables are local variables that are valid only once in the function in which they are declared
Static	Same as program	static	Static	Auto variables are local variables that are valid multiple times in the function in which they are declared
		extern	External	Global variables declared outside any function

Figure 4.87: Storage classes of variables.

Registers are fast storage locations inside CPUs. They provide the fastest way to access data, even surpassing RAM. However, the size of the registers is limited. Programmers nowadays seldom use register class themselves, because compilers will handle it automatically. Register class is often used for variables that are accessed frequently, such as loop variables.

Variables defined in functions are in the auto class by default unless otherwise specified. The value of an auto variable disappears when the function terminates. In its essence,
this happens because the system needs to reclaim storage units of the auto variable.

The value of a static variable is preserved after the function terminates. In other
words, the system will not reclaim its memory unit. Consequently, this value is still
available when the function is called again.

4.6.4 Masking mechanism 1: lifespan and scope of variables

4.6.4.1 Concept of scope

In a multiple-module structure, each child function possesses some internal data that
need not be accessed by other child functions. In the C language, the visibility of variables in a program is referred to as "scope." Figure 4.88 shows the definition of scope
and its rules.

Scope
A scope is the visibility of an object (such as a variable) in the code.

Rule of scope
Each function in a C program is an independent code block.
Code that constructs a function body is hidden from other parts of the program. It can't be accessed by statements (except the statement that calls the function) in other functions.

Figure 4.88: Scope and its rules.

4.6.4.2 Attributes of variables

Because information needs to be hidden from other functions in some cases, we need
to add an attribute to variables, which are carriers of information. Hence, attributes of
variables include data type and storage class.

Data type describes the size of memory a variable needed. The storage class indicates the lifespan of a variable in memory and its scope. Figure 4.89 shows the complete
syntax of variable declaration, which adds "storage class" in front of the data type.

Variable attribute	Data type	Data type indicates the size of a variable in the memory
	Storage class	Storage class indicates the life span of a variable in the memory and its scope

Complete syntax of variable declaration
storageClass   dataType   variableName

Figure 4.89: Variable attributes and declaration syntax.

Example 4.12 Usage of a local static variable

In Figure 4.90, we can see that there is an auto class local variable var and a static class local variable static_var in the child function varfunc(). After calling varfunc multiple times, we can make the following conclusion based on the execution result of the program: values of an auto class local variables disappear after the function returns, while values of a static class local variables are preserved even after the function returns.

```
01 #include"stdio.h"
02 void varfunc()
03 {
04    int var=0;   //Local variable
05    static int static_var=0; //Local static variable
06    printf("var=%d   ",var);
07    printf("static_var= %d\n",static_var);
08    var++;
09    static_var++;
10 }
11 int main(void)
12 {
13    int i;
14    for(i=0; i<3; i++)
15    {
16        printf("Iteration %d\n",i);
17        varfunc();
18    }
19    return 0;
20 }
```

Local variable: value is not accessible after the function terminates

Static variable: value is preserved after the function terminates

Program result:
Iteration 0 var=0 static_var= 0
Iteration 1 var=0 static_var= 1
Iteration 2 var=0 static_var= 2

Figure 4.90: Example of variable attributes and declaration syntax.

4.6.4.3 Local variables and global variables

Depending on the location of the definition, variables are divided into local variables and global variables. Figure 4.91 shows their definitions.

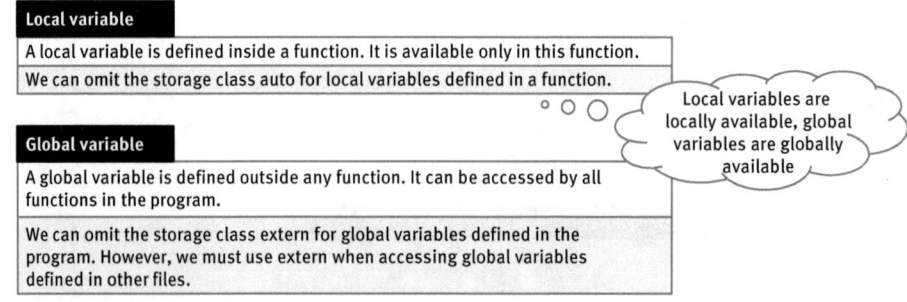

Local variable

A local variable is defined inside a function. It is available only in this function.

We can omit the storage class auto for local variables defined in a function.

Global variable

A global variable is defined outside any function. It can be accessed by all functions in the program.

We can omit the storage class extern for global variables defined in the program. However, we must use extern when accessing global variables defined in other files.

Local variables are locally available, global variables are globally available

Figure 4.91: Local variables and global variables.

Keyword extern can be added in front of variables and functions to indicate that their definitions are located in other files. Compilers will look for definitions in other modules upon seeing extern.

Example 4.13 Compute average score
In a competition, scores of players are truncated means of scores given by N judges. The highest and lowest scores are discarded, then the mean value of the rest is the final score of a player.

Analysis
1. Data structure design
Let scores be stored in array data[N]. As we need to access it in all processing steps, we can make it a global variable. To make testing easier, we can initialize it with initial values in our program, as shown in Figure 4.92.

Suppose there are N judges
We use a global variable data[N] to store scores

It is easier to test the program if we initialize the array

```
#define N 12                        //Number of judges
int data[N]= {86,96,92,88,93,94,89,88,91,90,87,91};
//Global variable of score array
```

Figure 4.92: Average score: data structure design.

2. Algorithm design

Based on the problem description, we can use the algorithm shown in Figure 4.93 to solve the problem.

Pseudo code	Refinement
Discard the minimum score Least	Find the minimum score Least, and replace it with 0
Discard the maximum score Largest	Find the maximum score Largest, and replace it with 0
Compute the average of data	Compute the average of data

We must replace the minimum with 0 before looking for the maximum.

Figure 4.93: Average score: algorithm design.

3. Function structure design
Because the score array data is global, that is, can be accessed by all functions, our function does not need to read data through parameters or return the updated data, as shown in Figure 4.94.

Functionality	Input information	Output information	Function header
Discard the minimum score Least	No (use global variable)	No (use global variable)	void Del_Least()
Discard the maximum score Largest	No (use global variable)	No (use global variable)	void Del_Largest()
Compute the average of data	No (use global variable)	Float value	float average()
Function name	Parameter list	Function type	

Figure 4.94: Average score: function design.

4. Code implementation

Figure 4.95 presents the implementations of each child function. Storage classes of local variables defined in these functions are omitted, so they are auto by default. To discard the minimum
score in data, we initialize Least with the first element of data and compare every other element
with Least until we find the minimum. After finding the minimum score, we update it to be 0. We
can discard the maximum score using the same way. Finally, we add all values in data together
and divide the sum by the number of judges minus 2 to obtain the mean.

```
//Discard the minimum value Least
void Del_Least()
{
    int Least,tag=0;
    Least=data[0];
    for(int i=0; i<N; i++)
    {
        if (Least>data[i])
        {
            Least=data [i];
            tag=i;
        }
    }
    printf("Least=%d \n",Least);
    data[tag]=0;
}
```

```
//Discard the maximum value Largest
void Del_Largest()
{
    int Largest,tag=0;
    Largest=data[0];
    for(int i=0; i<N; i++)
    {
        if(Largest<data[i])
        {
            Largest=data[i];
            tag=i;
        }
    }
    printf("Largest=%d \n",Largest);
    data[tag]=0;
}
```

```
//Compute the average of data
float average()
{
    float sum=0;
    for(int i=0; i<N; i++)
    {
        sum+=data[i];
    }
    return (sum/(N-2));
}
```

Figure 4.95: Average score: code implementation.

Figure 4.96 is a screenshot of part of the program, which includes function declarations, global
variables declarations, and the main function.

Note that on line 7, array data is declared outside the main function. Because it is declared
in the same file, extern can be omitted.

The three function calls are between line 63 and line 65. The first two functions are nonvalue-
returning functions, while the third is a value-returning function.

```
02 #define N 12   // The number of judges
04 void Del_Least();       // Discard the minimum value Least
05 void Del_Largest();     // Discard the maximum value Largest
06 float average();        // Compute the average of data
07 int data[N]= {86,96,92,88,93,94,89,88,91,90,87,91};
   // Score array as a global variable
55 int main(void)
56 {
57     float x;
58     for(int i=0; i<N; i++)
59     {
60         printf("%d ",data[i]);
61     }
62     printf("\n");
63     Del_Least();
64     Del_Largest();
65     x=average();
66     printf("average=%.2f\n",x);
68     return 0;
69 }
```

A global array defined
outside functions

86 96 92 88 93 94 89 88 91 90 87 91
Least=86
Largest=96
average=90.30

Figure 4.96: Partial code of average score problem.

Example 4.14 Scope of global variables
There are four functions in a program. Variables a, b, c, m and n are all global variables, but their scopes are different, as shown in Figure 4.97.

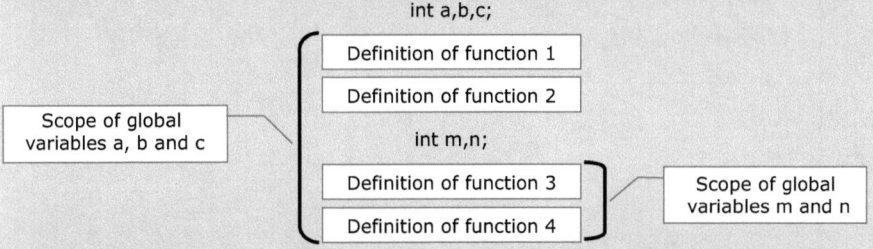

Figure 4.97: Scope of global variables.

Scopes of a, b and c are from function 1 to function 4, while m and n are only visible to the last two functions. In other words, function 3 and function 4 can use all these variables; function 1 and function 2 can only use a, b, and c.

We can conclude that scope of global variables depends on its location in the program.

Example 4.15 Local variables with duplicated names
Design a program where local variables have the same name, and examine their values at different moments.

Analysis
1. Program design
Let a and b be two local variables defined in main function and in child function sub. We assign values to them in both functions. The following code shows their values before and after calling sub:

```
1   #include <stdio.h>
2   void sub();
3
4   int main(void)
5   {
6     int a,b; //They are local variables in main function
7
8     a=3; b=4;
9     printf("main:a=%d,b=%d\n",a,b); //Print their values in main function
10    sub(); //Call sub, which assign new values to a and b
11    printf("main:a=%d,b=%d\n",a,b); // Print their values in main function
12    return 0;
13  }
14
15  void sub()
```

```
16 {
17   int a,b; //They are local variables in sub
18
19   a=6;  b=7;
20   printf("sub: a=%d,b=%d\n",a,b); //Print their values in sub
21 }
```

Program result:

main:a=3,b=4

sub: a=6,b=7

main:a=3,b=4

Note: Before calling sub, values of a and b are local variable values in main. When calling sub, their values are local variable values in sub. Values in main function are masked. After returning to main, values of a and b are once again local values in main.

2. Debugging

In Figure 4.98, we can see that addresses of a and b in main are 0x12ff7c and 0x12ff78 respectively.

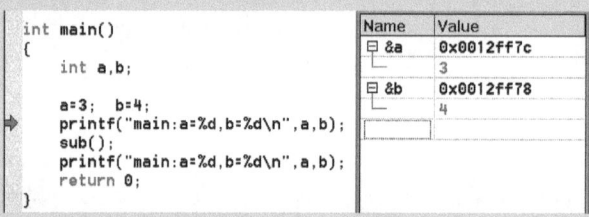

Figure 4.98: Local variables with duplicated names debugging step 1.

In Figure 4.99, the program has just entered sub function. Addresses of a and b have become CXX0069 error: variable needs stack frame. This error occurs because we have not allocated memory to the variables we wish to inspect. Are a and b in the window local variables in main function or in sub? Because the "execution arrow" points to the beginning of sub function, we can infer that local variables a and b in sub have not been declared. Thus, this is an error related to variables in sub function.

```
void  sub()
{
    int a,b;

    a=6;   b=7;
    printf("sub: a=%d,b=%d\n",a,b);
}
```

Name	Value
&a	CXX0069: Error:
&b	CXX0069: Error:

Figure 4.99: Local variables with duplicated names debugging step 2.

In Figure 4.100, it is clear that addresses of a and b are different from what we have seen before. Although these variables have the same name in main and sub, they are actually stored in different memory units

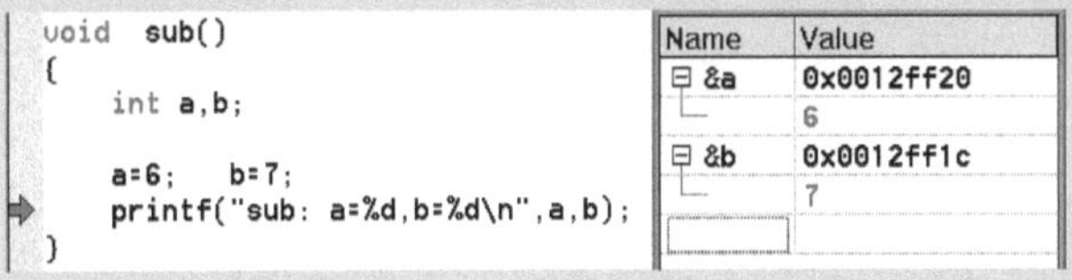

```
void  sub()
{
    int a,b;

    a=6;    b=7;
    printf("sub: a=%d,b=%d\n",a,b);
}
```

Name	Value
⊟ &a	0x0012ff20
└	6
⊟ &b	0x0012ff1c
└	7

Figure 4.100: Local variables with duplicated names debugging step 3.

.

In Figure 4.101, a and b are once again local variables visible in main function, after returning to main.

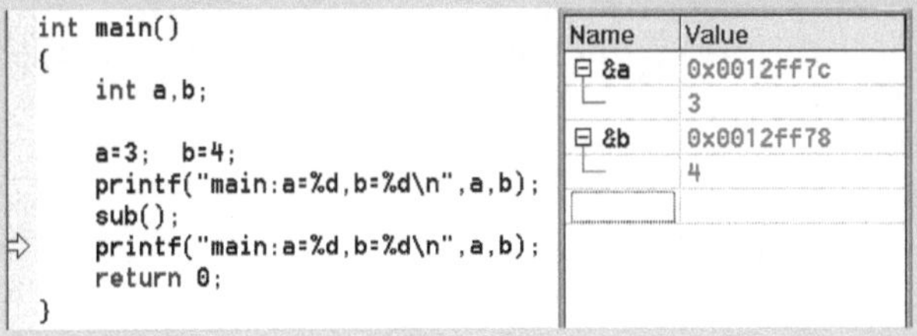

```
int main()
{
    int a,b;

    a=3;    b=4;
    printf("main:a=%d,b=%d\n",a,b);
    sub();
    printf("main:a=%d,b=%d\n",a,b);
    return 0;
}
```

Name	Value
⊟ &a	0x0012ff7c
└	3
⊟ &b	0x0012ff78
└	4

Figure 4.101: Local variables with duplicated names debugging step 4.

Example 4.16 Local variables and global variables with duplicated names
Examine scopes of a global variable and a local variable with the same name.

Analysis
Let a and b be two global variables. We also define local variables in child function max and in main with the same names. Both functions contain operations on variables a and b. The program is as follows.
1. Code implementation

```
1  #include <stdio.h>
2  int max(int a, int b);
3
4  int a=3,b=5; //Define a and b as global variables
5
6  int max(int a, int b) //a and b here are local variables
7  {
8    return (a>b ? a:b);
9  }
10
11 int main(void)
```

```
12 {
13   int a=8; //Define local variable a
14
15   printf( "max=%d\n", max(a,b)); //Use local variable a and global
16                                  //variable b as arguments
17   return 0;
18 }
```

Program result:

max=8

2. Debugging

In Figure 4.102, we have just entered main function. The Watch window displays both address and value of global variable b. CXX0069 error occurs for variable a because there is a global variable and a local variable with the same name.

In Figure 4.103, the address of local variable a is 0x12ff7c. No value has been assigned to it at this time.

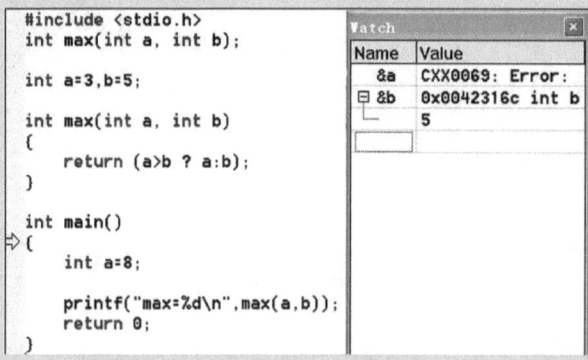

Figure 4.102: Local variables and global variables with duplicated names debugging step 1.

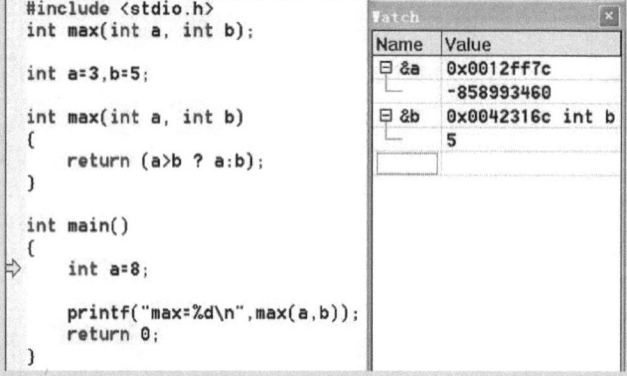

Figure 4.103: Local variables and global variables with duplicated names debugging step 2.

In Figure 4.104, we have assigned a value to a. In Figure 4.105, the address of local variable in child function is 0x12ff28, instead of 0x12ff7c of the local variable a in the main function. Global variable b is invisible in function max. Address of local variable b is 0x12ff2c.

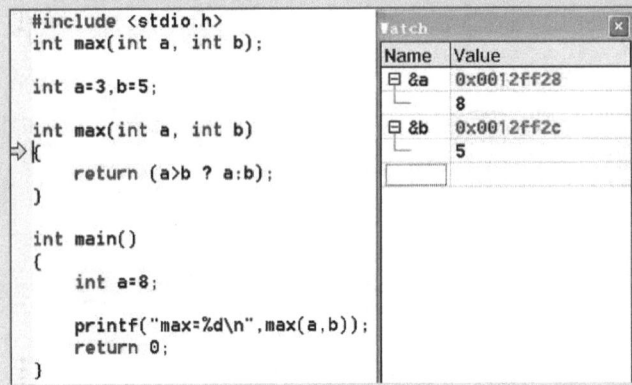

```c
#include <stdio.h>
int max(int a, int b);

int a=3,b=5;

int max(int a, int b)
{
    return (a>b ? a:b);
}

int main()
{
    int a=8;

    printf("max=%d\n",max(a,b));
    return 0;
}
```

Name	Value
&a	0x0012ff7c
	8
&b	0x0042316c int b
	5

Figure 4.104: Local variables and global variables with duplicated names debugging step 3.

```c
#include <stdio.h>
int max(int a, int b);

int a=3,b=5;

int max(int a, int b)
{
    return (a>b ? a:b);
}

int main()
{
    int a=8;

    printf("max=%d\n",max(a,b));
    return 0;
}
```

Name	Value
&a	0x0012ff28
	8
&b	0x0012ff2c
	5

Figure 4.105: Local variables and global variables with duplicated names debugging step 4.

Figure 4.106 shows addresses of a and b after returning to main. In Figure 4.107, the local variable a in the main function is changed to c. In this case, both global variables a and b are visible in the main function.

```
#include <stdio.h>
int max(int a, int b);

int a=3,b=5;

int max(int a, int b)
{
    return (a>b ? a:b);
}

int main()
{
    int a=8;

    printf("max=%d\n",max(a,b));
    return 0;
}
```

Watch ☒

Name	Value
⊟ &a	0x0012ff7c
∟	8
⊟ &b	0x0042316c int b
∟	5

Figure 4.106: Local variables and global variables with duplicated names debugging step 5.

```
#include <stdio.h>
int max(int a, int b);

int a=3,b=5;

int max(int a, int b)
{
    return (a>b ? a:b);
}

int main()
{
    int c=8;

    printf("max=%d\n",max(a,b));
    return 0;
}
```

Watch ☒

Name	Value
⊞ &a	0x00423168 int a
⊞ &b	0x0042316c int b
⊟ &c	0x0012ff7c
∟	8

Figure 4.107: Local variables and global variables with duplicated names debugging step 6.

Conclusion
It is recommended to use different names for local variables and global variables. Duplicated names will mask global variables, resulting in confusion.

3. Summary of local variables and global variables
Figure 4.108 summarizes the rules of local variables and global variables. If a local variable in a function has the same name as a global variable, the latter will be masked in this function. In other words, modifying the value of the local variable will not affect the global variable.

Rules

– Local variables are locally visible;

– Global variables are globally visible;

– If a local variable has the same name as a global variable, the local variable has the higher priority.

> If a local variable defined in a function has the same name as a global variable, the global variable is masked in this function.

Figure 4.108: Rules of local variables and global variables.

(1) Pros of global variables:
  - Easier data transmission: it is easier and more convenient to transmit data among functions through referencing global variables than using parameters and return statements.
  - Higher execution efficiency: with global variables, fewer parameters are necessary for functions. This reduces the cost of calling functions, thus improving the execution efficiency.
(2) Cons of global variables:
  - Worse universality: using global variables affects encapsulation and universality of functions.
  - Worse readability: it is more difficult to debug programs with global variables because it is hard to figure out which function makes the global data wrong.
  - More memory: memory is not allocated to global variables as needed. Instead, memory allocated to global variables will not be reclaimed until the program terminates.

As a conclusion, it is not recommended to use global variables unless the performance of programs is of vital significance.

4.6.5 Masking mechanism 2: scope of functions

In the "shared resources" example, we mentioned that functions also have the "availability" attribute. If a function defined in a source file can only be called by functions in the same file, it is called an internal function; if functions in other files can call it as well, it is called an external function.

To identify internal functions and external functions, we use two keywords of storage classes: static and extern. When used for this purpose, they are merely identifiers and are no longer indicators of storage classes.

Figure 4.109 shows how to define an internal function and how to declare and define an external function. The scope of an internal function is restricted to its source file, while the scope of an external function is the entire program. If we omit

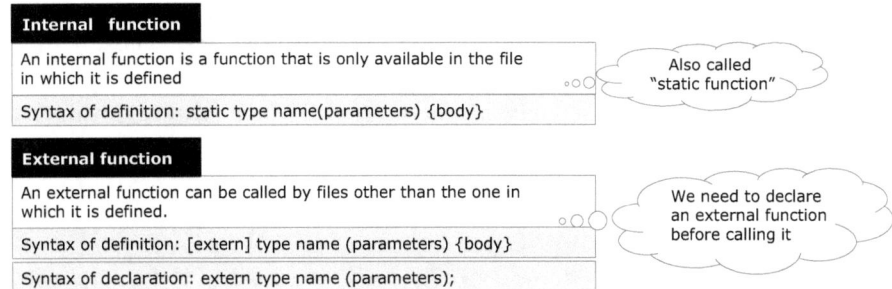

Figure 4.109: Syntax of internal and external functions.

extern when defining a function, it will be implicitly defined as an external function. In files that need to call external functions, it is necessary to use the extern keyword to indicate that the function called is external.

The merit of using internal functions is that one need not worry about whether his/her functions have the same names as functions written by others when various people write functions of the program because it does not matter in this case. For example, buildings in Mr. Brown's universities are named by letters in the alphabet. His friend works for another university, whose buildings are also named by letters. However, teachers and students in Mr. Brown's university will never confuse their buildings with buildings in other universities.

Example 4.17 A program with multiple files

Suppose a C program consists of three source files, namely test file 1 1.cpp, test file 2 2.cpp, and test file 3 3.cpp, which are shown in Figures 4.110–4.112, respectively. The main function is located in test file 1. These figures contain definitions of functions, comments on external declarations, and comments on global variables. Please refer to Appendix A for how to add multiple files to a project in the IDE.

Differences between a declaration and a definition: when a function or a variable is declared, no physical memory is allocated. The purpose of declaration is to make sure compilers can compile the program. When a function or a variable is defined, it is stored in physical memory space. A function or a variable can be declared multiple times, but it can only be defined once.

How do we reference global variables in multiple files? Suppose we define a global variable in a file, we need to use extern keyword in this file to make it accessible in other files. On the other hand, if we define a static global variable, it is only accessible in this file, instead of other files.

4.6.6 Masking mechanism 3: restriction on shared data

We have seen in previous examples that data are often shared among functions. Functions can access the same data objects at different stages in different ways. Sometimes, an unintentional operation may change the data, which is not what we expect to see.

Our goal is to make data accessible by multiple functions and to ensure they cannot be arbitrarily modified. To do this, we can use const keyword to define data in parameters as constant. const is a keyword of C which prevents a variable from being modified. Using const can partially enhance security and robustness of programs.

Test file1.cpp:

```
01 #include <stdio.h>
02 extern  int reset(void);    //Declare reset as an external function
03 extern  int next(void);     // Declare next as an external function
04 extern  int last(void);     // Declare last as an external function
05 extern  int news(int);i     // Declare news as an external function
06
07 int i=1;                     //Define global variable i
08 int main()
09 {
10    int i, j;                 //Define local variables I and j
11    i=reset();
12    for (j=1; j<4; j++)
13    {
14       printf("%d    %d    ",i, j );
15       printf("%d    ",next());
16       printf("%d    ",last());
17       printf("%d\n",news(i+j));
18    }
19    return 0;
20 }
```

These are function declarations.
Their definitions are not in this file

Whether a variable is global depends
the location at which it is defined
(outside all functions)

Global variable i and local variable i
are not store in the same memory unit

Call an external function

```
Program result
1   1   2   3    7
1   2   4   5   10
1   3   6   7   14
```

Figure 4.110: A program with multiple files 1.

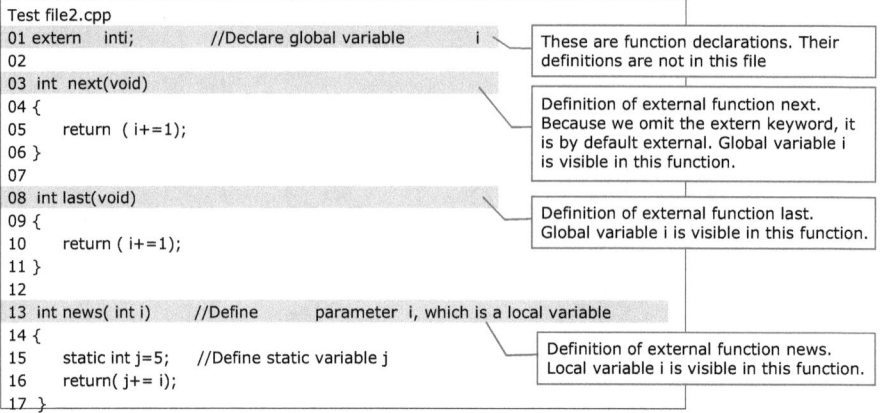

Figure 4.111: A program with multiple files 2.

```
Test file3.cpp
01 extern int i;          //Declare local variable    i
02 int  reset(void)
03 {
04     return ( i);
05 }
```

Definition of external function reset.
Global variable i is visible in this function.

Figure 4.112: A program with multiple files 3.

Example 4.18 Using const to prevent array from being modified

```
1   //Example of const
2   #include <stdio.h>
3   #define SIZE 3
4   void modify( const int a[] ); //Function prototype
5   int b[SIZE]; //Globa variable used to store updated array
6
7   //Program starts from main
8   int main(void)
9   {
10     int a[SIZE] = { 3, 2, 1 }; //Initialization
11     int i; //Counter
```

```
12
13   modify(a); //Function
14   printf( "\n Array a after calling modify:" );
15   for( i = 0 ; i < SIZE ; i++ )
16   {
17     printf( "%3d", a[i] ); //Print the array after function call
18   }
19
20   printf( "\n Array b after calling modify:" );
21   for( i = 0 ; i < SIZE ; i++ ) //Print updated array
22   {
23     printf( "%3d", b[i] );
24   }
25   return 0;
26 }
27
28 //Fetch values from array a, process them and store results in array b
29 void modify(const int a[])
30 {
31   int i; //Counter
32   for(i=0;i<SIZE;i++)
33   {
34   //a[i]=a[i]*2; Compilation error if we attempt to modify a
35     b[i]=a[i]*2;
36   }
37 }
```

Program result:
Array a after calling modify: 3 2 1
Array b after calling modify: 6 4 2

4.7 Recursion

To iterate is human, to recurse, divine. – L. Peter Deutsch

4.7.1 Case study

Last weekend, Mr. Brown took Daniel to a family reunion. Since Daniel had never attended a reunion, Mr. Brown introduced him to the four other kids in the room.

Five kids then sat together. When Mr. Brown asked about their age, the first kid A said, "I am 2 years older than B on my left." B decided to do the same and said, "I am 2 years older than C on my left as well." C imitated, "I am 2 years older than D on my left." D said, "I am 2 years older than Daniel." When it came to Daniel, he answered honestly that he was 10.

Mr. Brown burst into laughter and asked, "This is fun. How will you solve this problem?"

"I am 10, so D is $10+2 = 12$, C is $12+2 = 14$, B is $14+2 = 16$, and A is $16+2 = 18$." Daniel answered quickly. "Well done!" said Mr. Brown, "Can any of you generalize a formula?" B reckoned that this was a recursive relation. A, who had been learning to program, said that this could be easily implemented by a loop, as shown in Figure 4.113.

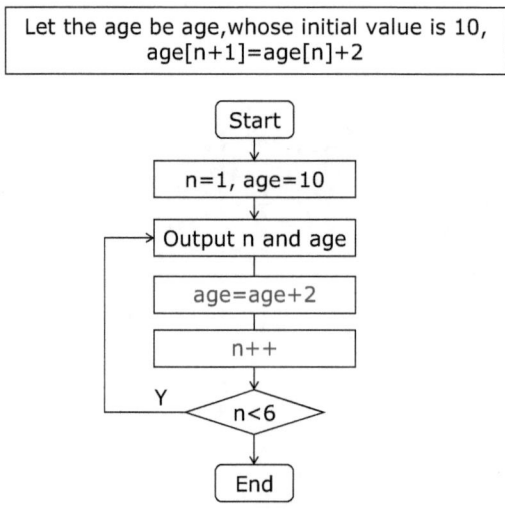

Figure 4.113: Age computing solution 1.

"Is this the only method?" asked Mr. Brown.

A thought for a little while and said, "We can write the formula in another way to simulate the process of Mr. Brown asking about our age." He then changed age $[n+1] = age[n]+2$ to $age[n] = age[n-1]+2$. See Figure 4.114 for the derivation. "We can't directly compute the age of the fifth person age[5], but it is related to the age of the fourth person age[4]. Although we can't compute age[4] either, we can use age[3] to express it. Repeating this process, with the new recursive formula, we get to age[1], which is already known, and then compute all the way back using the original recursive formula."

Prof. Brown applauded him and asked what "n decreases by 1 proactively" and "n increases by 1 passively" meant.

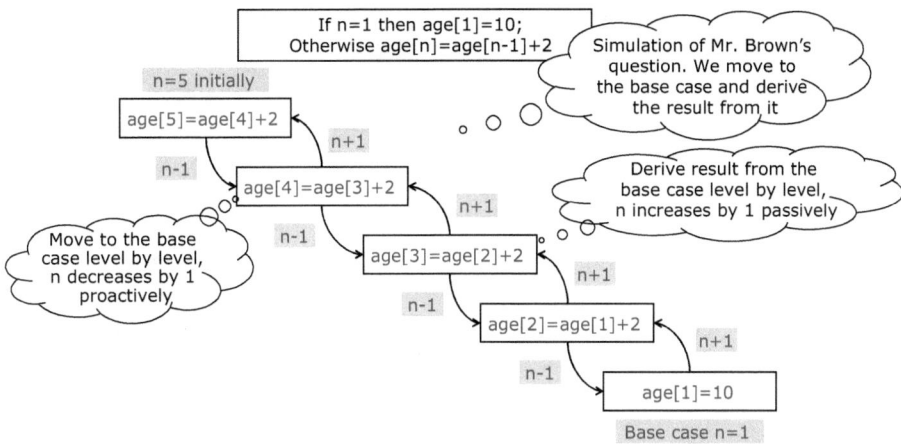

Figure 4.114: Age computing solution 2.

"'n decreases by 1 proactively' means that we scale down the problem. Here we are decreasing the number of people n. What to decrease and how to decrease should be determined by us. 'n increases by 1 passively' refers to the upscaling when we come back from the base case. In this case, it is the increment of n. How larger the scale of a previous problem is than the current one is determined during the downscaling. Here we are increasing by 1" answered A.

Then Mr. Brown asked if anyone could draw the execution flow of this solution. Receiving no response after a while, he drew the graph himself, as shown in Figure 4.115.

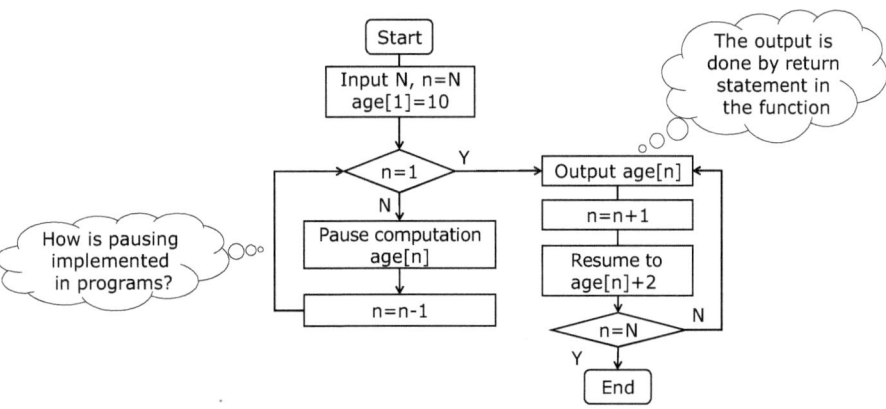

Figure 4.115: Flowchart of age computing solution 2.

"But how do we pause in a program? You can say that in words, but statements in programs are executed one after another," asked A.

"Calling another function," Mr. Brown responded, "pauses the caller and executes the callee. Since we have a 'pause' in the second method, it is only possible through calling another function."

"It is hard to see the correspondence between the change of n and age[n], though," argued A.

"In this case, let the child function be int age(int n), then the calling relation between the main function and age can be represented by the chart in Figure 4.116. Please note that there are multiple 'pausing points.' The age()'s in the gray area are all pausing points."

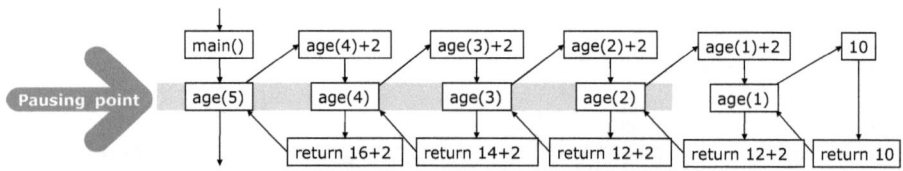

Figure 4.116: Schematic of the execution process of age computing solution 2.

According to this figure, we can write the code shown in Figure 4.117.

```
01 #include <stdio.h>
02 int age(int n)
03 {
04     if (n==1) return(10);        // Base case
05     else return (age(n-1)+2);    // Move to the base case and derive result from it
06 }
07                          ┌── age calls itself
08 int main(void)
09 {
10     printf("%d",age(5));
11     return 0;
12 }
```

Figure 4.117: Code implementation of age computing solution 2.

On line 5, we can see that the function age() is calling itself but with a smaller argument.

"It is like looking into the mirror." A said, "Standing between two mirrors, you can see many images of yourself, each smaller than another." See Figure 4.118 for an illustration of this metaphor.

Figure 4.118: Cat in the mirror in the mirror.

When a computation process calls itself directly (or indirectly), we call it a recursive process. If the description of an object contains itself, or it defines itself, then we call such an object a recursive object.

A recursive process is a round-trip process. As the scale of a problem becomes smaller and smaller, there is an endpoint at which the scale can no longer be decreased. Then we start from the endpoint and return to where we started along the original path.

4.7.2 Concept of recursion

We have roughly talked about recursion in the age example. Now we are going to define recursion formally.

4.7.2.1 Definition of recursion

Term explanation Recursion
In mathematics and computer science, recursion refers to a function that uses itself in its definition.

The basic idea of recursion is to convert a large-scale problem into similar subproblems of a smaller scale. Because we use the same function to solve similar problems of different scales, the function may need to call itself. Moreover, the function must have a termination condition to

obtain results. Otherwise, it will call itself infinitely. As a result, a recursive process must contain two key elements:
- Base case: the most straightforward instance of the problem, which can be solved without recursion.
- Recursive case: an instance of the problem that can be solved through solving more straightforward instances.

4.7.2.2 Type of recursion

If statements in a function call the function again, either directly or indirectly, we call it a recursive call of functions. Figure 4.119 shows an example of a direct recursive call, in which a statement inside func calls func again.

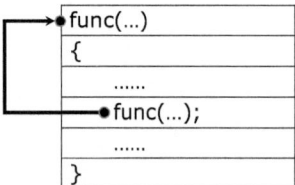

Figure 4.119: Direct recursive call.

Figure 4.120 shows an example of an indirect recursive call. A statement in func1 calls func2, then a statement in func2 calls func1.

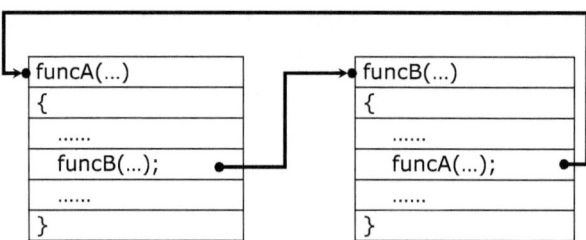

Figure 4.120: Indirect recursive call.

Compared with nested function calls, recursive calls are special cases of nested calls: the functions being called are exactly the caller. Both direct and indirect recursive calls lead to a loop of function calls. If there is no base case, the program will end in a situation that is similar to infinite loops.

4.7.2.3 Comparison of recursion and iteration

Recursion and iteration are based on program control structures: iteration uses a loop structure, while recursion uses branch structure. Both of them involve using loops: iteration uses loops explicitly, while recursion uses loops through repeated function calls. Both recursion and iteration require a termination condition: iteration terminates when the loop condition evaluates to false, while recursion stops upon the base case.

4.7.3 Example of recursion

Example 4.19 Computing factorial using recursion

$$n! = \begin{cases} 1, & n = 0 \\ n \cdot (n-1)!, & n > 0 \end{cases}$$

Analysis

1. Algorithm description

The process of computing factorial is as follows:
- n! can be computed by n * (n–1)!, so it suffices to compute (n–1)!;
- Similarly, it suffices to compute (n–2)! to obtain (n–1)!;
- In this way, n gets smaller and smaller. When n = 1, 1! is something we already know;
- then we trace back to compute 2!;
- and 3!;
- finally, we trace back to n!;

2. Code implementation

```c
#include "stdio.h"
float fac(int n);
//fac computes n!
float fac(int n)
{
  float f;
  if (n<0) printf("Error!\n"); //Input is invalid when n<0
  if (n==0||n==1) return 1; //Base case
  return n*fac(n-1); //n! = n * (n-1)!
}
int main(void)
{
  printf("%f",fac (4) );
  return 0 ;
}
```

3. Execution process of recursion

The calling process of the recursion is shown in Figure 1.121. Unlike nested calls of multiple functions, all child functions are the same in recursion.

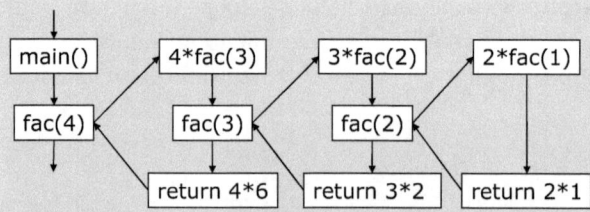

Figure 4.121: Process of main function calling fac.

4. Efficiency analysis of recursion

One major drawback of recursive functions is system cost. Whenever a function is called, the system has to allocate stack space to store parameter information. If we use recursive functions, the system needs a large amount of memory for stack spaces. If we use a large integer as the argument of fac, the system may crash in the worst case.

Good programming habit

We should use as few recursive calls as possible. Recursive calls, especially indirect ones, make programs less readable. Besides, recursive calls require many system resources. Furthermore, it is hard to test programs with recursive calls. Hence, we should avoid unnecessary recursive calls, unless using them simplifies some algorithms or functions.

In recursion, we derive an unknown value by stepping backward; in loops, we derive an unknown value by repeating the same process, which is a forward process.

Program reading exercise Compute nth item of Fibonacci sequence

Analysis

We can derive the key elements of recursion based on the recursion relation of the Fibonacci sequence.
- Base cases: $fab(1) = 1$, $fab(2) = 1$.
- Recursive cases: $fab(n) = fab(n-2) + fab(n-1)$.

The code implementation is as follows:

```
int fab(int n)
{
 if(n==1 || n==2) return 1;
 else return (fab(n-1)+fab(n-2));
}
```

Program reading exercise Computing 1+2+ . . .+n with recursion

Analysis
- Base case: $f(1) = 1$
- Recursive cases: $f(n) = n + f(n-1)$。

The code implementation is as follows:

```
int fn(int n)
{
```

```
if (n < 1) return 0; //Exception handling
else if (n == 1) return 1;
else return (n + fn(n - 1));
}
```

Program reading exercise Finding the maximum element in an array with recursion

Analysis
Let the array be arr[] with length len.
- Base case: when len = 1, the maximum is the first element arr[0].
- Recursive cases: the maximum is the larger of arr[0] and the maximum of the array starting from the second element.

The three key elements of the function are as follows:
- Functionality: finding the maximum element in an array.
- Input: array address and array length.
- Output: the maximum

The critical step in the algorithm: the maximum of the array starting from the second element is max(arr + 1, len-1).
    The code implementation is as follows:

```c
#include <stdio.h>
int max(int arr[], int len)
{
  if(1 == len) //Only one element
  {
  return arr[0];
  }
  int a = arr[0]; //The first element
  int b = max(arr + 1, len - 1); //Maximum of the array starting from
the second element
  return a > b? a : b;
}
int main(void)
{
  int a[] = {1,2,3,4,5,6,7,8,9,10};
  printf("Maximum: %d\n", max(a, sizeof(a) / sizeof(a[0])));
  return 0;
}
```

4.8 Summary

1. Three syntaxes related to functions: declaration, definition, and call.
2. Three key elements of function design: input, output, and functionality determine function structures.
   (1) Element 1: function name describes the functionality.
   (2) Element 2: input determines numbers and types of parameters.
   (3) Element 3: output determines function type.

3. Three ways of data transmission between functions: return statement, argument, and global variables.
   (1) return statement: only one value can be passed from the called function to the caller.
   (2) argument: caller passes arguments to the function by address or by value.
   (3) global variables: accessible by all functions.

Figure 4.122 shows the main contents of this chapter and their relations.

We can divide large-scale problems into independent modules, each implemented by a child function,
Repeated tasks can also be implemented as a code segment.
The manufacturer defines how a function is implemented,
While the users call functions to complete tasks.
Input, output, and functionality define the structure of functions,
The function name describes the functionality,
Data to be processed are put into the parameter list,
The type of output determines the function type.
Child functions need to communicate with the caller,
The caller uses them through function calls,
Actual data are transmitted as arguments,
We use pass by value for single datum and pass by reference for a group of data,
Return, arguments, and global variables are used for the other direction,

We should select one depending on the problem we want to solve.
We use storage classes to identify the lifespans of variables.
The scope of variables and functions may vary.
Local variables are restricted to functions,
While global variables are visible to everyone.

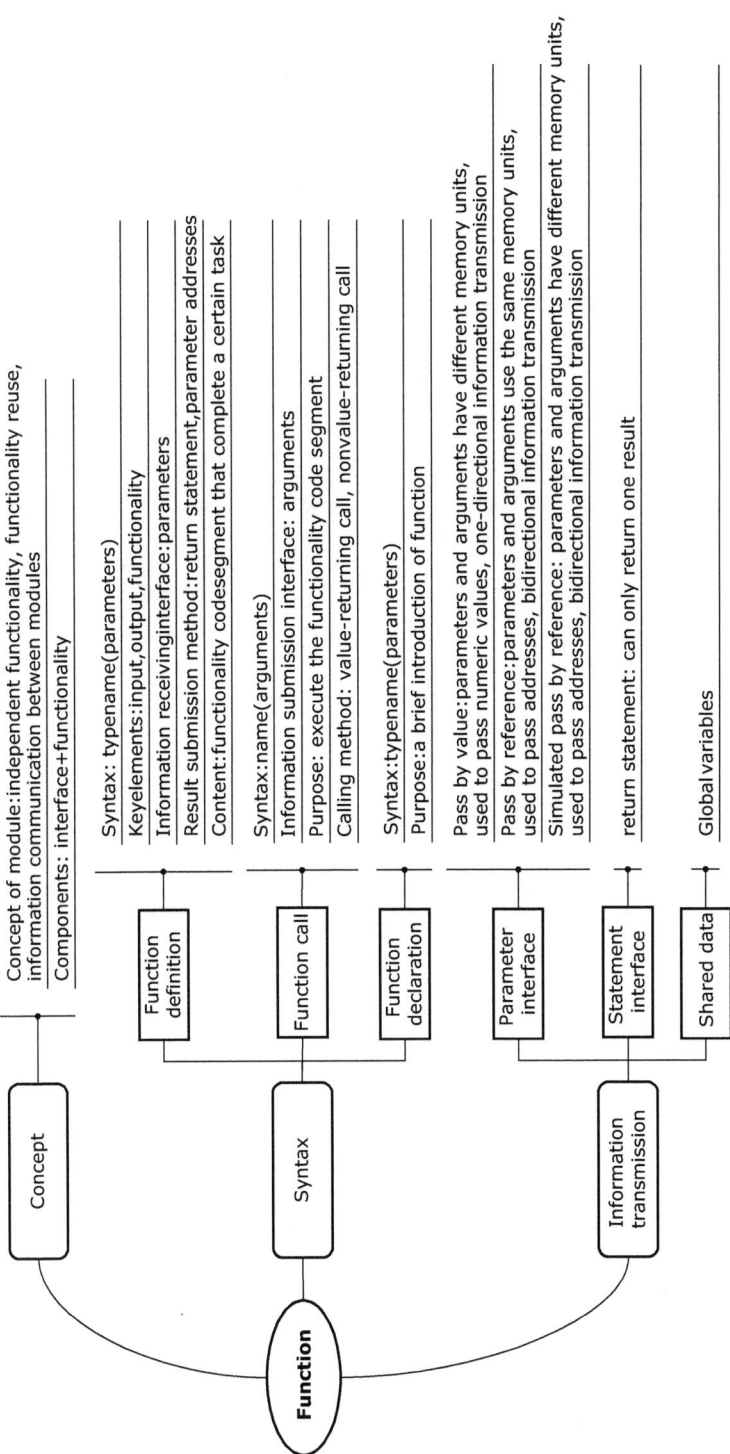

Figure 4.122: Concepts related to functions and their relations.

4.9 Exercises

4.9.1 Multiple-choice questions

1. [Concept of functions]

    Which of the following statements is correct? ( )
    A) A function in a C program can call or be called by any other functions in the same program.
    B) The location of the main function in a C program is fixed.
    C) We cannot define another function in a function.
    D) Every C program file has to have a main function.

2. [Pass by value]

```
void fun( int a, int b )
{
 int t;
 t=a; a=b; b=t;
}
int main(void)
{
 int c[10]={1,2,3,4,5,6,7,8,9,0}, I;
 for (i=0;i<10; i+=2) fun(c[i], c[i+1]);
 for (i=0;i<10; i++) printf("%d,", c[i]);
 printf("\n");
 return 0;
}
```

What is the output of the program above? ( )
    A) 1,2,3,4,5,6,7,8,9,0,
    B) 2,1,4,3,6,5,8,7,0,
    C) 0,9,8,7,6,5,4,3,2,1,
    D) 0,1,2,3,4,5,6,7,8,9,

3. [Pass by reference]

```
#define N 4
void fun(int a[][N], int b[],int n)
{
 int i;
 for(i=0;i<n;i++) b[i] = a[i][i];
}
```

```
int main(void)
{
  int x[][N]={{1,2,3},{4}, {5,6,7,8},{9,10}}, y[N], i;
  fun(x, y, N);
  for (i=0;i<N; i++) printf("%d,", y[i]);
  printf("\n");
  return 0;
}
```

What is the output of the program above? ( )

A) 1,0,7,0,   B) 1,2,3,4,   C) 1,4,5,9,   D) 3,4,8,10,

4. [A void function that accepts pointers]

```
void swap(char *x, char *y)
{
  char t;
  t=*x; *x=*y; *y=t;
}
int main(void)
{
  char *s1="abc", *s2="123";
  swap(s1,s2);
  printf("%s,%s\n",s1,s2);
  return 0;
}
```

What is the output of the program above? ( )

A) 321,cba   B) abc,123   C) 123,abc   D) 1bc,a23

5. [Recursion]

```
#include <stdio.h>
void my_put()
{
  char ch;
  ch = getchar();
  if (ch != 'C') my_put();
  putchar(ch);
}
```

```
int main(void )
{
 my_put();
 return 0;
}
```

Suppose input is: ABC <Return>, what is the output of the program above? ( )
A) ABC   B) CBA   C) AB   D) ABCC

6. [Global variables]
Which of the following statements is wrong about global variables? ( )
   A) The scope of a global variable starts from its definition and ends at the end of the source file.
   B) A global variable is one that can be defined at any position outside functions.
   C) We can restrict the scope of a global variable using the extern keyword.
   D) The lifespan of a global variable is the entire execution process of the program.

7. [Scope]
Which two storage classes include variables that only take up memory units when being used? ( )
   A) auto and static
   B) extern and register
   C) auto and register
   D) static and register

8. [Concept of storage classes]
Which of the following statements is wrong? ( )
   A) The system does not assign a specific initial value automatically to an auto variable defined in C functions.
   B) We can define variables in each compound statement in the same function. The scope of these functions is restricted to the compound statement.
   C) Suppose we define a static variable with an initial value in a function. Every time the function is called, the variable will be assigned an initial value.
   D) Parameters of a function cannot be static variables.

4.9.2 Fill in the tables

Fill in the tables in Figures 4.123–4.127 based on programs in each problem.

1. [Pass by value]

```
double fun( double x, int y)
{
  int i;
  double z;
  for(i=1, z=x; i<y;i++) z= z* x;
  return z;
}
```

```
  for(i=1, z=x; i<y;i++) z= z* x;
    return z;
}
```

i	1	2	3	...	y-1	y
z	x^2					
Functionality of fun						

Figure 4.123: Functions: fill in the table question 1.

2. [Pass by value with one-dimensional array]

Fill in the table in Figure 4.124 based on the program below.

```
#define N 5
void sub( int n, int uu[])
{
  int t;
  t=uu[n-1]+uu[n];
  uu[n]=t;
}
int main(void)
{
  int i, aa[N]={1,2,3,4,5};
  for(i=1; i<N; i++) sub(i,aa);
  for(i=0; i<N; i++) printf("%d_",aa[i]);
```

```
  printf("\n");
  return 0;
}
```

n	1	2	3	4
(1) aa[n]	1,2,3,4,5,6			
t	1+2			
(2) uu[n]	uu[1]=3			
Program output:				

Figure 4.124: Functions: fill in the table question 2.

3. [Pass by reference with two-dimensional array]

```
#define N 3
#define M 3
select(int a[N][M], int *n)
{
  int i,j,row=1,colum=1;
  for(i=0;i<N;i++)
  for(j=0;j<M;j++)
  if(a[i][j]>a[row][colum]){ row=i; colum=j; }
  *n= row;
  return ( a[row][colum]);
}
int main(void)
{
  int a[N][M]={9,11,23,6,1,15,9,17,20},max,n;
  max=select(a,&n);
  printf("max=%d,line=  %d\n",max,n);
  return 0;
}
```

		0	1	2
a[N][M]	0	9	11	23
	1			
	2			

i		0			1			2		
j		0	1	2	0	1	2	0	1	2
a[row][colum]	9									
row	0									
colum	0									

Figure 4.125: Functions: fill in the table question 3.

4. [Purpose of using static variables in functions]

```
int ff(int n)
{
  static int f=1;//——②
  f=f*n;
  return f;
}
int main(void)
{
  int i;
  for(i=1;i<=5;i++) //——①
  printf("ff=%d\n",ff(i));//——③
  return 0;
}
```

①i	1	2	3	4	5
②f	5				
③ff(i)					

Figure 4.126: Functions: fill in the table question 4.

5. [Pass by reference with structure pointer]

```
#include<string.h>
#define N 3
struct stu
{
  int ID;
  char name[10];
  int age;
};
int fun(struct stu *p)
{
  p->ID+=201700; //——①
  return (strlen(p->name)); //——②
}
int main(void)
{
  struct stu students[N]=
  {{1, "Zhang",20},
  {2, "Wang", 19},
  {3, "Zhao", 18}};
  int len;
  for(int i=0;i<N; i++ )
  {
   len=fun(students+i);
   printf("Name %d has length %d\n",students[i].ID,len);
  }
  return 0;
}
```

i	0	1	2
p	&students[0]		
①p->ID			
②p->name			

Figure 4.127: Functions: fill in the table question 5.

4.9.3 Programming exercises

1. Please write a program that reads two integers from keyboard input and outputs the one's digit of the larger and the smaller to the second power.

2. Sequence A is defined as follows:

```
A(1)=1,
A(2)=1/(1+A(1)),
A(3)=1/(1+A(2)),
. . . . . .
A(n)=1/(1+A(n-1))。
```

Please write a function that computes the nth item of the sequence.

3. Please write a function that reads a string from the main function, computes, and outputs its length.

4. Suppose users type in multiple words in the console. Words are separated by spaces. The '#' sign is used to indicate the end of input. Please write a function that converts the first letter of each word into uppercase. The input is handled in the main function.

5. Please write a program that: reads a nonzero integer n from keyboard input, computes the sum of each digit of n and outputs the sum if the sum is a one-digit number. If the sum has multiple digits, the program should repeat the above process until the sum has single digit.

   For instance, the conversion process of n = 456 is as follows:

   $4 + 5 + 6 = 15$

   $1 + 5 = 6$

   The output is 6

6. Please write a program that reverses n input numbers using pointers.

   Requirements:

   (1) The program should have only one main function.

   (2) Write a child function for reversing numbers. Input/output of data should be done in the main function.

5 Files: operations on external data

!

5.1 Introduction

A had been learning the C language and thought it was interesting. He was eager to solve some practical problems with what he had learned.

One day, the class president asked him to compute the ranking of average grades of his classmates in the midterm exam. He then used his programming knowledge and completed the task quickly. His program asked users to input grade information of every student in the class and then printed the sorted average grades onto the screen.

However, the class president complained to A after trying the program, "I have given you access to the electronic version of grades, but you didn't use it. Instead, your program asked me to input every grade on the keyboard. That was too tedious. Also, your program merely printed the result on the screen. I had nothing left after I closed it. Our school requires us to submit an electronic record. Your program is not user-friendly enough, and I can't use it."

The class president's complaint made A speechless. "How can a program read electronic records of grades? We have only learned to read keyboard input. Besides, how do I save the result to a data file?" he thought to himself.

To sum up, A's question is: how should we fetch data from and save the result to persistent storage automatically with programs?

As we all know, the purpose of programming is to process data as needed to complete specific tasks. The data processing flow consists of data input, data processing, and result output. Execution and testing of programs also involve data input and

https://doi.org/10.1515/9783110692303-005

output. Hence, data input/output is a critical part of programming. By analyzing the data input/output process, we know the following things:

(1) Data to be processed are either created by programmers in programs (in this case, programs can only process these data) or input by users during program execution (users need to reenter the data in each execution of the program).

(2) Results of programs are output to screen instead of saving permanently.

It was these two features of normal data input/output that made A's program not user friendly. In practice, we often have the following needs regarding data processing:

(1) Input: the amount of input data is large; the input data are always the same.

(2) Output: we need to inspect the results frequently; the output is too much to be displayed on a screen without scrolling.

In these cases, we can save these data for easier inspection or repeated use. To save data permanently in computers, we store them into external memory. Operating systems manage data in the external memory in the form of files. As a result, it is necessary to learn file operations to complete programming tasks quickly and flexibly.

5.2 Concept of files

A file is an ordered set of correlated data. The name of a file is called filename. We have encountered files in previous chapters many times. For example, we have mentioned source files, object files, executable files, and library files (header files). Files are a persistent form of data, and they make data sharing possible.

Depending on how data are stored, files in C can be divided into binary files and text files.

5.2.1 Binary files

Binary files, as the name indicates, store data in binary codes. For example, integer 5678 is stored as 00010110 00101110, which takes up 2 bytes in memory (the hexadecimal form of 5678 is 0x162E).

Although we can view binary files on screen, their contents are often garbled characters because they are mostly nontext characters.

5.2.2 Text files

Text files are also called as ASCII code files. Each character in such a file is stored as a 1-byte ASCII code on disks. For example, Figure 5.1 shows the storage format of number 5678.

Binary form	0011,0101	0011,0110	0011,0111	0011,1000
Character	'5'	'6'	'7'	'8'

Figure 5.1: Representation of characters in text files.

ASCII code files can be displayed as characters on screen. For example, source files are also ASCII code files. We can read them because they are displayed as characters.

The difference between text files and binary files is that: text files are constructed by characters, while binary files are constructed by bits. Note that both of them are handled as "stream files" in the C language.

Term explanation
Stream files: C treats files as "data streams," which are sequences of consecutive bytes with no breaks. Such a structure is called a "stream file structure," in which each byte is accessible. A termination mark exists at the end of a file, which is similar to the string termination mark.

We do not bother figuring out data characteristics, types, and storage formats when processing stream files. We merely access data in bytes. The analysis and processing of data are left to be done by other programs. As a result, this file structure is more flexible and can better utilize storage space.

5.2.3 File termination mark and end-of-file checking function

(1) End-of-file (EOF) is the file termination mark. EOF is an integer symbolic constant defined in header file as <stdio.h>, whose value is usually −1. It is worth noting that EOF is only used for text files because −1 is also a valid character in binary files.
(2) feof function is a function in the standard library, which is used to determine whether we have reached the end of a file. It works for both binary files and text files.

Knowledge ABC About EOF   i
Using constant EOF instead of −1 enhances the portability of programs. ANSI C standard emphasizes that EOF must be a negative integer (but not necessarily −1). As a result, the value of EOF varies in different systems. The input method of EOF also depends on which system we are using, as shown in Figure 5.2.

System	Input method of EOF
UNIX-like	<return> <ctrl-d>
Windows	<ctrl-z>

Figure 5.2: Input methods of EOF in different systems.

Files are regarded as data streams in C. There is also a file termination mark and a function to determine whether we have reached the end. In practice, the internal pointer of a file points the beginning of the stream by default when users open it with programs. As users execute operations, the pointer can move to other positions in the stream. Finally, we check whether the pointer points to the end of the file to make sure we have read the entire file.

5.3 Operation flow of files

By now, we should have had a basic idea of files. How do we operate files in practice? Files are usually stored in an external medium (like disks) and brought to internal memory when needed. We call the process of data moving to memory from disks "read" and the process of data moving to disks from memory "write."

In operating systems, each file is identified by a unique filename. Computers use filenames to read and write a file.

When we look for data in a file on the disk by ourselves, we must find the file using its name, read data from it, and close it. We use the same steps to operate files with programs. The three fundamental steps of accessing files with programs are:
(1) Opening the file
(2) Processing the file
(3) Closing the file

File operations that are available in programs are as follows:
(1) Creating and saving a new file on disks
(2) Opening an existing file
(3) Reading from and writing to a file

We know that C has no input/output statements. The input/output of data is done by calling library functions. In a broader sense, the operating system regards all input/output devices connected to the computer as files. Input and output are then similar to reading from and writing to a disk file. We usually define the monitor as the standard output file. Displaying information on the screen is then outputting to the standard output file. Functions like printf and putchar are all in this category. The keyboard is often regarded as the standard input file. Typing in information through the keyboard is then inputting data from the standard input file. Functions like scanf and getchar are examples of such input.

ANSI defines standard input/output functions and uses them to read and write files. Readers can refer to Appendix C of Volume 1 for details of these functions. We shall analyze some of the most common library functions in subsequent sections.

5.4 Data communication between internal and external memory

As shown in the file operation flow introduced in Section 5.3, the internal memory and the external memory must communicate with each other to implement read and write operations of files. Ideally, we want such communication to be done simultaneously. However, the reality is cruel. Different components of computer work at different speeds, so tasks are often completed at different times. To solve this problem, we introduce the buffer system into the communication between the internal and the external memory.

Knowledge ABC Buffers

Because CPU and RAM work at high speed and external memory (like disks or CDs) works slower, the computer has to wait for the external memory before it can proceed to subsequent work. This speed mismatch affects the CPU's performance terribly. As a result, the "buffer" technology was introduced to solve the problem, as shown in Figure 5.3.

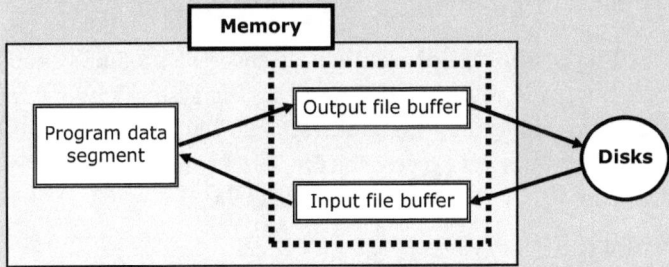

Figure 5.3: Buffer system.

A buffer is a block of storage space in the internal memory, allocated and managed by the system upon opening a file. The size of a buffer depends on the version of C. It usually is multiples of 512 bytes.

When writing data to files in the external memory, we do not directly write to the external memory. Instead, we write to the buffer. When the buffer is full or the file is closed, data in it are automatically written to the external memory. It is the same for reading from a file. At first, only one block of data is read into the buffer. When we read the data, we first look for them in the buffer. If they exist, we simply fetch them from the buffer. Otherwise, we search for them in the external memory. After finding the data we want, we read the data block in which they are located in the buffer. Buffers can effectively reduce external memory accesses.

Reading and writing using buffers can better utilize disks. Standard C also uses a buffer system.

When using a buffer system, the system creates a buffer for each file opened. Operations on files then become operations on buffers.

For programmers' convenience, ANSI C defines a structure for information related to file buffers (such as filename corresponding to the buffer, operations allowed on the file, size of the buffer, and location of the data being accessed in the buffer). We can obtain information about file buffers by accessing this structure variable. The type of this structure is FILE, which is defined in stdio.h (note: based on what we have learned about header files, we must include this header file when using FILE to operate files).

The information contained in FILE type is as follows:

```
typedef struct _iobuf
{
  char* _ptr;   //Points to the first unread character in the buffer
  int _cnt;   //Number of remaining unread characters
  char* _base;   //Points to a character array, namely buffer of this file
  int _flag;   //A flag for some properties of the file
  int _file;   // Used to obtain file description. We can obtain file descriptor of the
  file using fileno function
  int _charbuf;   //Single byte buffer. If the buffer is single-byte, _base is then invalid
  int _bufsiz;   //Size of the buffer
  char* _tmpfname;   //Temporary file name
} FILE;
```

Whenever a file is opened successfully, the operating system creates a FILE variable for the file, allocates memory, and returns a pointer to it. The system stores information about the file and the buffer into this FILE variable. Our program can use the pointer to obtain file information and access the file, as shown in Figure 5.4.

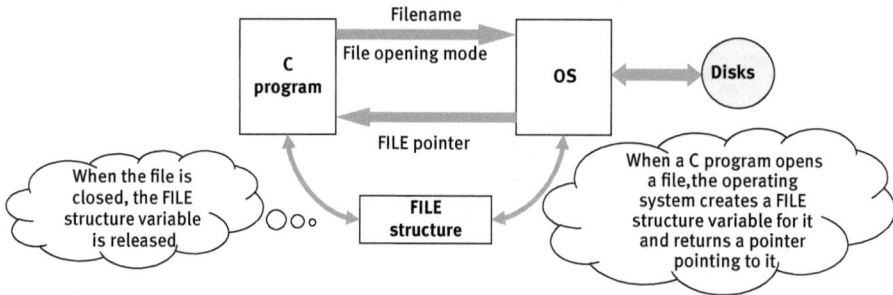

Figure 5.4: File operations.

After the file is closed, the variable is freed.

As long as we have this file pointer, we can use file operation functions provided by the system to operate the file without knowing details of the buffer. File operating code is thus easier to write. Now we are going to study how to operate files.

5.5 Operations on files using programs

We have introduced in Section 5.3 that programs operate file following three steps: opening files, reading files, and closing files. There are corresponding library functions for all these steps in ANSI C. We shall analyze these functions in the following sections.

5.5.1 Opening files

The library function for opening file is fopen, whose detailed information is as follows:
 − Prototype: FILE fopen (char *filename, char *mode)
 − Functionality: allocate a file buffer for a file in the memory.
 − Parameters:
   filename: a string that contains the path and the name of the file to be opened
   mode: a string indicating the mode of file opening.
 − Return value: file pointer (NULL indicates that the file was not opened because an exception happened)

Note: Beginners often ignore exceptions when programming. They often think that the file is opened after calling fopen function and uses the returned file pointer directly. However, this is problematic. fopen do not always open files successfully. Invalid filenames or not enough access privileges can lead to an exception in fopen. It is recommended to check if the file is successfully opened after calling fopen. To be more specific, we should check whether the returned file pointer is NULL before actually accessing the file.

The mode parameter of fopen determines the mode in which the file is opened. There are multiple modes, whose values and meaning are shown in Figure 5.5.

Note:
(1) The opened file can be either text file or binary file.
(2) A text file is represented by "t" (optional), while a binary file is represented by "b."

Programming error
As shown in Figure 5.5, mode "w" always checks whether the file exists first, regardless of the file being a text file or a binary file. If the file exists, the function deletes the existing file and create a new one. As a result, we should be careful when using it. If we want to preserve the original contents in the file, we should not use mode "w" because it will delete the contents without any warnings.

File opening mode		Meaning
Read-only	"r"	Open a text file in read-only mode, fail if the file doesn't exist
	"rb"	Open a binary file in read-only mode, fail if the file doesn't exist
Write-only	"w"	Open a text file in write-only mode, create a new file if the file doesn't exist, delete and create a new one if it exists
	"wb"	Open a binary file in write-only mode, create a new file if the file doesn't exist, delete and create a new one if it exists
Read/write	"r+"	Open a text file in read/write mode, fail if the file doesn't exist
	"rb+"	Open a binary file in read/write mode, fail if the file doesn't exist
	"w+"	Open a text file in read/write mode, create a new file if the file doesn't exist, delete and create a new one if it exists
	"wb+"	Open a binary file in read/write mode, create a new file if the file doesn't exist, delete and create a new one if it exists
	"a+"	Open a text file in read/write mode, create a new file if the file doesn't exist, append to the file if it exists
	"ab+"	Open a binary file in read/write mode, create a new file if the file doesn't exist, append to the file if it exists
Append	"a"	Append data to the end of a text file, create a new file if the file doesn't exist, append if it exists
	"ab"	Append data to the end of a binary file, create a new file if the file doesn't exist, append if it exists

Figure 5.5: File opening mode.

Although it is not grammatically wrong, using the wrong file opening mode can lead to logic execution errors. For example, when we use write mode "w" to open a file instead of using update mode "r+," the file contents will be deleted. In conclusion, we must determine the correct file opening mode before accessing files.

Knowledge ABC File paths

A path is the sequence of directories we need to visit when searching for a file on the disks. There are absolute paths and relative paths. An absolute path starts from the drive letter. It is a complete description of the location of a file. A relative path is a location relative to the target location. It starts with the current directory.

The string that can uniquely identify a disk file is

```
Drive letter:\Path\Filename.Extension
```

Ex. 1: We are looking for the file c:\windows\system\config. If we are currently in the directory c:\windows\, then the relative path is system\config, and the absolute path is c:\windows\system\config.

Ex. 2:

```
fp=fopen("a1.txt","r");
```

This is a relative path with no path information. In this case, file a1.txt is in the current directory (note: the current directory refers to the directory of the project which contains this program).

```
fp=fopen("d:\\qyc\\a1.txt","r");
```

This is an absolute path and file a1.txt is located in the directory qyc in D drive.

Note: we use "\\" instead of "\" because "\" should be escaped in strings.

5.5.2 Reading and writing

Unlike opening files, there are many cases of reading and writing, so people created a series of library functions for them, as shown in Figures 5.6 and 5.7.

Functionality	sFunction	Standard I/O counterpart
Read/write character	int fgetc(FILE *fp)	getchar()
	int fputc(int ch,FILE *fp)	putchar()
Read/write formatted data	int fscanf(FILE *fp,char *format,arg_list)	scanf()
	Int fprintf(FILE *fp,char*format,arg_list)	printf()

Figure 5.6: File reading and writing functions 1.

Note: We read from and write to the current position of a file. The current position is the position currently pointed to by the data read/write pointer. When a file is opened, the pointer points to the beginning of the file; after we read or write a byte successfully, the pointer moves forward automatically (moves to the next byte).

Functionality	Function	Parameters
Read/write strings	char * fgets (char *str, int num,FILE * fp)	num: the number of characters being read str: the address of the char array
	int fputs (char *str, FILE * fp)	
Read/write data blocks	int fread (void *buf,int size,int count,FILE * fp)	count: the number of data entries size: the length of a data entry buf: the address of the buffer
	int fwrite (void *buf,int size,int count,FILE * fp)	

Figure 5.7: File reading and writing functions 2.

Now we are going to introduce these functions through examples.

Example 5.1 Example of files 1
Read and display characters in file file.txt.

Analysis

```
1   //Read characters one by one from file
2   #include <stdio.h>
3   #include <stdlib.h>
4
5   int main(void)
6   {
7     char ch;
8     FILE *fp;   //Define a FILE pointer fp
9     fp=fopen("file.txt","r");   //Open text file file.txt in read-only mode
10    if (fp==NULL)   //Failed to open the file
11    {
12      printf("cannot open this file\n");
13      exit(0);   //Call library function exit to terminate the program
14    }
15    ch=fgetc(fp);   //Read a character and assign it to ch
16    while(ch!=EOF)   //Check whether we have reached the end, equivalent to
      (!feof(fp)) in this case
17    {
18      putchar(ch);   //Output the character
19      ch=fgetc(fp);   //Read a character and assign it to ch
20    }
21    fclose(fp);   //Close the file
22    return 0;
23 }
```

Note: We use while(ch!=EOF) to determine whether we have reached the end of the file on line 16. This statement only works for files opened as text file. If we open a file in binary modes, then we should use !feof(fp). Otherwise, we may wrongly consider a file to be completely read when seeing value "–1."

Term explanation
exit function: exit is declared in <stdlib.h>. It is used to terminate a program forcibly. When there are input errors or the program cannot open a file, we can use this function to end the program. The parameter of exit is passed to some operating systems so that other programs can use it.

exit(0) means the program exits normally. In contrast, exit(1) indicates an exception (there must be an exception as long as the argument is not 0, but we recommend using the macro EXIT_FAILURE defined in stdlib.h to indicate the reason of exception. The macro is defined as 1 in the header file).

Knowledge ABC What are the differences between exit() and return in C?
Exit function is used to exit the program and return to the operating system, while a return statement merely returns to the caller from the function currently being executed. If we use return in the main function, then the program terminates after return is executed and returns to the operating system. In this case, the return statement is equivalent to exit. However, one merit of exit is that we can call it in other functions and use a search program to look for these calls.

Example 5.2 Example of files 2
Write the specified string into a file, and read the string from the file into an array.

Analysis
```
1  //Write the specified string into a file
2  #include <stdio.h>
3  char *s="I am a student";   //Specify the string to be written
4  int main(void)
5  {
6    char a[100];
7    FILE *fp;   //Define file pointer fp
8    int n=strlen(s);   //Compute length of s
9
10   //Open text file f1.txt in write mode
11   if ((fp=fopen("f1.txt","w"))!=NULL)
12   {
13     fputs(s,fp);   //Write string pointed by s into file pointed by fp
14   }
15   fclose(fp);   //Close the file pointed to by fp
16
17   //Open text file f1.txt in read-only mode
18   fp = fopen("f1.txt","r");
19   fgets(a, n+1, fp);   //Read contents in file pointed to by fp into array a
20   printf("%s\n",a);   //Print a
21   fclose(fp);   //Close the file pointed to by fp
22   return 0;
23 }
```

Note: fgets(a, n+1, fp) on line 19 reads a string and stores it in array a. a is a character array defined earlier. n+1 instructs the program to read n characters from the file pointed to by fp and store them into a. These n characters are precisely string s. Because a string must be terminated with "\0", we use n+1 instead of n.

Think and discuss Is it necessary to check the result of file opening function?
Discussion: good programmers try their best to consider all possible error cases when programming. In this example, we use a short string s for the convenience of demonstration, so it is fine to write line 19. However, string s may be an extremely long string in practice. In this case, we have to consider whether a is large enough to store the string in order to avoid out-of-bound errors. Besides, we did not check the result of fopen when opening the file in read-only mode, which is a risk in the program.

Example 5.3 Example of files 3
Write formatted data onto the disk, and then read the contents from the file and display them on the screen.

Analysis

```
1   //Write data block into a file
2   #include "stdio.h"
3   #include "stdlib.h"
4
5   struct student   //Define the structure
6   {
7     char name[15];
8     char num[6];
9     float score[2];
10  } stu;
11  int main(void)
12  {
13    FILE *fp1;
14    int i;
15
16    fp1=fopen("test.txt","wb");
17    if( fp1 == NULL)   //Open file in binary write-only mode
18    {
19      printf("cannot open file");
20      exit(0);
21    }
22    printf("input data:\n");
23    for( i=0;i<2;i++)
24    {
25    //Input a row of record
26      scanf("%s%s%f%f",
27      stu.name,stu.num,&stu.score[0],&stu.score[1]);
28    //Write data block into the file, one row at a time
29      fwrite(&stu,sizeof(stu),1,fp1);
30    }
31    fclose(fp1);
32
33    //Open the file again in binary read-only mode
34    if((fp1=fopen("test.txt","rb"))==NULL)
35    {
36      printf("cannot open file");
37      exit(0);
38    }
39    printf("output from file:\n");
40    for (i=0;i<2;i++)
41    {
42      fread(&stu,sizeof(stu),1,fp1);   //Read block from the file
```

```
43   printf("%s %s %7.2f %7.2f\n",   //Display on the screen
44   stu.name,stu.num,stu.score[0],stu.score[1]);
45 }
46 fclose(fp1);
47 return 0;
48 }
```

Program result:
```
input data:
xiaowang  j001 87.5 98.4
xiaoli    j002 99.5 89.6
output from file:
xiaowang  j001 87.50 98.40
xiaoli    j002 99.50 89.60
```

Programming error
After writing content into a file, we may need to read the file later. Sometimes, we may see garbled characters in the file. This is due to the inconsistency between the format we used when writing to the file and the format of the file operating function. In the example above, if we change line 29 into fprintf, there will be an output error.

5.5.3 Closing files

There is an old saying which goes "Timely return of a loan makes it easier to borrow a second time." We should return things we borrow from others quickly. If our credit is good, then people are likely to help us when we need to borrow the second time. In programs, dynamically allocated resources should follow this rule as well. Otherwise, a memory leak may happen. In the worst cases, it will lead to results beyond our expectations. The FILE pointer in file operations is also a resource. We obtain it by successfully calling fopen function. As a result, we have to return this resource after using the file. The return here refers to closing the file. Readers may have noticed that we always call fclose function after we are done with the file in previous examples. fclose is the function we use to close files. It is defined as follows:
– Prototype: int fclose(FILE *fp)
– Functionality: it closes the file pointed to by the file pointer, handles the data in the buffer, and releases the buffer eventually.
– Output: if an exception happens, the function returns a nonzero value; otherwise, it returns 0.

Note: we should close a file in time after we use it. Otherwise, data may get lost. Data are not written into the file until the buffer is full. If we terminate the program when the buffer is not yet full, data in the buffer will be discarded.

Example 5.4 Example of files 4

Analysis
```
1   //Write 10 record into data.txt
2   #include <stdio.h>
3   int main(void)
4   {
5     FILE *fp; //FILE is the file type
6     int i;
7     int x;
8
9     fp=fopen("data.txt","w"); //Open data.txt in text write mode 'w'
10
11    for(i=1;i<=10; i++)
12    {
13      scanf("%d",&x);
14      fprintf(fp,"%d",x); //Output x into the file pointed to by fp
15    }
16    fclose(fp); //Close the file
17    return 0;
18  }
```

Program result: we can find the newly created file data.txt in the directory of our project after the program terminates. We will see the 10 records read from keyboard input in it.

5.5.4 Random access

We have introduced the three steps of file operations in previous examples. Readers may have noticed that we could only read the file from the very beginning to the very end, one byte after another. Is it always acceptable in practice?

Apparently, such a rigid method is not always suitable in real life. Suppose we have a file of student information, the records are stored in the order of student ID. We wish to quickly locate a row using its index like we do with arrays. It is obvious that we cannot do this with sequential access. In response to our needs, C provides the fseek function that can relocate the file pointer. It is defined as follows:
- Prototype: fseek(FILE pointer, offset, beginning location)
- Functionality: relocate the file pointer. It moves the pointer by "offset" bytes from the "beginning location" (Value of the beginning location: beginning of the file is represented by SEEK_SET, whose value is 0; current location is represented by SEEK_CUR, whose value is 1; the end of the file is represented by SEEK_END, whose value is 2).
- Return value: 0 is returned upon success, while –1 is returned upon a failure.

Example 5.5 Example of files 5
We have records of students in the file stu_list.txt. Write a program that reads the data of the second student.

Analysis
Code implementation:

```
1  //Read from specified location in a file: random access of files
2  #include "stdio.h"
3  #include "stdlib.h"
4
5  struct stu      //Structure of student information
6  {
7    char name[10];
8    int num;
9    int age;
10   char addr[15];
11 } boy,*qPtr;      //Define a structure variable boy and a structure pointer qPtr
12
13 int main(void)
14 {
15   FILE *fp;
16   char ch;
17   int i=1;      //Skip the first i rows
18   qPtr = &boy; //qPtr points to the beginning address of boy
19
20   if ((fp=fopen("stu_list.txt","rb"))==NULL)
21   {
22     printf("Cannot open file!");
23     exit(0);
24   }
25 //Relocate the pointer to the beginning of the file
26   rewind(fp);
27 //Move the pointer by (i*structureSize) bytes
28   fseek(fp,i*sizeof(struct stu),0);
29 //Read the current row from the file, and store into address pointed to by qPtr
30   fread(qPtr, sizeof(struct stu),1,fp);
31   printf("%st%5d %7d %sn", qPtr->name,
32   qPtr->num, qPtr->age, qPtr->addr);
33   fclose(fp);
34   return 0;
35 }
```

Note: To make this program work, the file has to be written and opened in binary mode. For example, we use fopen("stu_list.txt","rb") on line 20. Only in this case are the contents of the file binary data stored sequentially. Besides, we cannot use fseek(fp,i*sizeof(struct stu),0) on line 28 to move the pointer if the file is not accessed in binary mode.

5.6 Discussion on file reading and writing functions

When checking the file after we write to it, sometimes we find nothing but garbled characters. Sometimes, the binary data we read from a file are not what we have expected. What happened behind the scene?

We shall briefly discuss this problem by introducing several combinations of file open modes and file operating functions.

5.6.1 Case 1: fprintf and fscanf

In this case, we read the file data.txt in binary mode, use fprintf to write data, and use fscanf to read data from it.

```
1   //File reading and writing
2   #include <stdio.h>
3   #include <stdlib.h>
4
5   int main(void)
6   {
7     FILE *fp;   //FILE is the file type
8     int i;
9     int x;
10    int b=0;
11    char ch;
12    fp=fopen("data.txt","wb");   //Open data.txt in "wb" mode
13
14    if (fp==NULL)                //Fail to open
15    {
16      printf("1:cannot open this file\n");
17      exit(0);   //Terminate the program with exit
18    }
19  //***Using fprintf to write data****
20    for( i=1; i< 7; i++ )
21    {
22      scanf("%d",&x);
23      fprintf( fp,"%d",x );   //Output x to the file pointed to by fp
24    }
25    fclose(fp);   //Close the file
26
27    fp=fopen("data.txt","rb");   //Open text file data.txt in read-only mode
28    if (fp==NULL)                //Fail to open
29    {
30      printf("2:cannot open this file\n");
31      exit(0);   //Terminate the program with exit
32    }
```

```
33
```

```
34 //******* Using fscanf to read data ********
35  fscanf(fp,"%d",&x);   //Read an int value into x
36  while (!feof(fp) )  //Check whether the file ends
37  {
38    printf("%d ",x);
39    fscanf(fp,"%d",&x);
40  }
```

```
41  fclose(fp);  //Close the file
42  return 0;
43 }
```

Program result:
Input: 2 3 4 5 6 7
Output: 2 3 4 5 6 7

If we open data.txt manually, we will find that the contents are 234567, which can be displayed normally. If we use text editors like EditPlus to open the file in hexadecimal mode, we will find the bytes being ASCII values "32 0A 33 0A 34 0A 35 0A 36 0A 37 0A."

5.6.2 Case 2: fwrite and fread

In this case, we read the file data.txt in binary mode, use fwrite to write data and use fread to read data from it. The code implementation can be obtained by replacing the code segments in squares in case 1 with the code segments as follows:

```
//*** Using fwrite to write data ****
for( i=1; i<7; i++ ) //Write 6 int values into the file
{
  scanf("%d",&x);
  fwrite(&x,sizeof(int),1,fp); //Output x to the file pointed to by fp
}
```

```
//*** Using fscanf to read data ****
for( i=1; i<7; i++ )
{
  fread(&b,sizeof(int),1,fp);
  printf("b=%x\n",b);
}
```

Program result:
Input:
2 3 4 5 6 7
Output:
b=2
b=3
b=4

```
b=5
b=6
b=7
```

If we manually open data.txt in the operating system, we will find garbled characters in it. Opening the file in hexadecimal mode, we will see the bytes "02 00 00 00 03 00 00 00 04 00 00 00 05 00 00 0006 00 00 00 07 00 00 00." These bytes are precisely the binary byte stream of integers 2 to 7, in which each integer takes up 4 bytes. They are displayed as garbled characters because they are not stored as ASCII values. In the program above, however, the numbers can be correctly read because the fread function does read them as binary integer data.

5.6.3 Case 3: fprintf and fscanf

In this case, we read the file data.txt in binary mode, use fprintf to write data, and use fscanf to read data from it. The code implementation is the same as in the first case, except the code in the second square is replaced with the following statements:

```
//***Using fread to read data****
for( i=1; i<7; i++ )
{
  fread(&b,sizeof(int),1,fp);
  printf("b=%x\n",b);
}
```

```
Program result:
Input:
2 3 4 5 6 7
Output:
b=0a330a32
b=0a350a34
b=0a370a36
b=0a370a36
b=0a370a36
b=0a370a36
```

If we open data.txt manually, we will see numbers 234567 displayed correctly, as they were in case 1. However, the numbers displayed are different from the inputs because we use fread to read them as binary integers.

5.6.4 Case 4: fwrite and fscanf

In this case, we read the file data.txt in binary mode, use fwrite to write data and use fread to read data from it. The code implementation is the same as in the first case, except the code in the first square is replaced with the following statements:

```
//***Using fwrite to write data****
for( i=1; i<7; i++ )     //Write 6 integers into the file
{
  scanf("%d",&x);
  fwrite(&x,sizeof(int),1,fp);     //Output x into the file pointed to by fp
}
```

```
Program result:
Input:
2 3 4 5 6 7
Output:
Output "7" infinitely
```

The contents of the data.txt file are the same as in case 2. Nonetheless, the program runs into exception because fscanf cannot recognize binary bit stream correctly.

We used binary mode to read data in all 4 cases. Will the results be different if we use text mode? It is not hard to infer from our analysis in these cases that the results will be similar. Interested readers may try text mode in these 4 cases on their own.

Conclusion
When operating files, we should use matching functions for reading and writing. In this case, we can guarantee the data are correctly recognized regardless of using binary mode or text mode.

Whether the generated file can be displayed correctly is determined by the writing function, instead of the file opening mode. When we use fprintf, the file contains ASCII values, which can be displayed normally. When we use fwrite, data are written into the file as a binary bit stream. Whether they are normally displayed depends on whether they are valid ASCII values. We see garbled characters because they are not ASCII values in most cases.

5.7 Debugging and I/O redirection

After designing an algorithm and writing the code, we need to use test data to test the program in the debugging environment. Because we often find bugs in our programs, we need to rerun them and input test data repeatedly. In programs with many input data, it takes a long time to type on the keyboard. Is there a better way to do this? Here come files to save the day.

We can put input data in a file and read them with file reading functions and write results into specified files with file writing functions. Based on the characteristics of the test data, we should select suitable file operation functions. There are two code templates for this process.

5.7.1 Code template 1 Using fscanf and fprintf

```
#include <stdio.h>
int main(void)
{
  FILE *fp1, *fp2;
  fp1=fopen("data.in","r"); //Open input file data.in in read-only mode
  fp2=fopen("data.out","w"); //Open output file data.out in write-only mode
//We process our data here. Note that we should fscanf to read and fprintf to print
  fclose(fp1);
  fclose(fp2);
  return 0;
}
```

This program simply uses basic file operations we have learned. Now we are going to see a program using freopen function.

5.7.2 Using freopen function

- Prototype: FILE *freopen(const char*path,const char *mode,FILE *stream);
- Parameters:
  path: it is the filename used to store the custom input/output file name;
  mode: file opening mode, which is the same as in fopen;
  stream: a file, which is usually the standard stream files.
- Functionality: redirect the standard stream file to the file specified by path.
- Return value: the function returns a pointer to the file specified by path upon success; otherwise, it returns NULL (we rarely use its return value though).

Knowledge ABC Standard stream files

When we run a C program, the operating system opens three files and provide the program with pointers to them. These three file pointers are standard input stdin, standard output stdout, and standard error stderr. They are declared in <stdio.h>.

Stdin: standard input stream. It outputs to screen by default.

Stdout: standard output stream. It outputs to screen by default.

Stderr: standard error stream. It outputs to screen by default.

When a file is not accessible due to some reason, debugging information has to be printed to the end of output with stderr. This is acceptable when we print to screen. However, it is not acceptable when we write to files or write to other programs through pipes (a pipe is a buffer of fixed size). Output to stderr will be displayed on the screen even if we redirect the standard output.

5.7.3 Code template 2 Using freopen function

```
#include <stdio.h>
int main(void)
{
 freopen("data.in", "r", stdin);//Redirect input from keyboard to data.in
 freopen("data.out", "w", stdout);
 //Redirect output from screen to data.out

 //The data processing code remains the same

 fclose(stdin);
 fclose(stdout);
 return 0;
}
```

Using freopen is as simple as using fprintf and fscanf. Besides, we do not need to modify our code because we use input/output redirection, which is more convenient than the first template. Here is an example of redirection.

Example 5.6 Debugging the program that calculates a+b
(1) Keyboard input case.

```
1   #include <stdio.h>
2   int main(void)
3   {
4     int a,b;
5
6     while(scanf("%d %d",&a,&b)!=EOF)
7     {
8       printf("%d\n",a+b);
9     }
10    return 0;
11 }
```

Program result:

5 6

11

^Z

(2) Read data from in.txt and write result to out.txt.

```
1   #include <stdio.h>
2   int main(void)
3   {
4   int a,b;
5   //Redirect input, read data from in.txt under Debug directory of the
6   //current project
7   freopen("debug\\in.txt","r",stdin);
8   //Redirect output, write data to out.txt under Debug directory of the
9   //current project
10   freopen("debug\\out.txt","w",stdout);
11   while (scanf("%d %d",&a,&b)!=EOF)
12   {
13    printf("%d\n",a+b);
14   }
15   fclose(stdin); //Close the file
16   fclose(stdout); //Close the file
17   return 0;
18 }
```

Note:

(1) We read input data from in.txt under Debug directory of the current project. Before running the program, we need to save our data "5 6" in in.txt (beware of the space between 5 and 6. Although the program is still valid if we omit the space, the result will not be what we expected. Interested readers can try the program without the space, examine the result and analyze why the result is different using what we have learned).

(2) We save output data to out.txt under Debug directory of the current project. After running the program, we will find a out.txt file under Debug directory, which contains the number "11."

5.8 Summary

Figure 5.8 shows main contents of this chapter and relations between them.

Files are persistent form of data,
We can save data in binary form or text form.
Programs follow three steps to operate files:
Namely opening, reading/writing, and closing.
To open a file, we need its path and filename;
To operate a file, we use library functions;
After operating the file, we must close it.

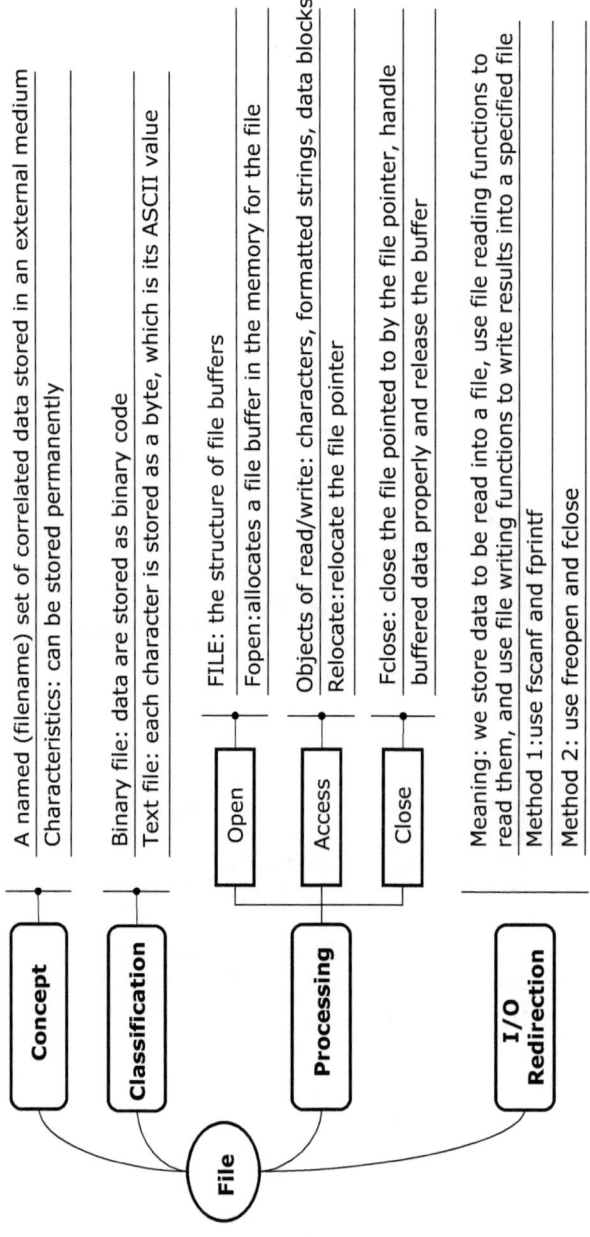

Figure 5.8: Relations between fundamental concepts related to files.

5.9 Exercises

5.9.1 Multiple-choice questions

1. [Concept of files]
   Which of the following statements is correct about files in C? ( )
   A) File is constructed by a series of data. It must be a binary file.
   B) File is constructed by a series of structures. It could be a binary file or a text file.
   C) File is constructed by a series of data. It could be a binary file or a text file.
   D) File is constructed by a series of characters. It must be a text file.

2. [Opening a file]
   Suppose we have the following code segment:

```
FILE *fp;
if( (fp=fopen("test.txt", "w")) == NULL)
        {printf("Failed to open file!");
        exit(0);}
else
        printf("File opened successfully!");
```

   If the file test.txt does not exist and there is no other exception, which of the following statements is wrong? ( )
   A) The output is "Failed to open file!"
   B) The output is "File opened successfully!"
   C) The system will create a file with the specified name.
   D) The system will create a text file for write operation.

3. [fprintf]
   Suppose we have the following program.

```
#include <stdio.h>
int main(void)
{    FILE *f;
     f=fopen("filea.txt","w");
     fprintf(f,"abc");
     fclose(f);
     return 0;
}
```

   If the content of filea.txt was originally: hello, then then content of it after running the program above will be ( )
   A) abclo   B) abc   C) helloabc   D) abchello

4. [fseek and rewind]

   Suppose we have the following program.

```c
#include <stdio.h>
int main(void)
{
    FILE *fp;
    int i, a[6]={1,2,3,4,5,6},k;
    fp = fopen("data.dat", "w+");
    fprintf(fp, "%d\n", a[0]);
    for (i=1; i<6; i++)
    {
        fseek(fp, 0L, 0);
        fscanf(fp, "%d", &k);
        fseek(fp, 0L, 0);
        fprintf(fp, "%d\n", a[i]+k);
    }
    rewind(fp);
    fscanf(fp, "%d", &k);
    fclose(fp);
    printf("%d\n", k);
    return 0;
}
```

   What is the output of this program? ( )
   A) 21   B) 6   C) 123456   D) 11

5. [End of file]

   Suppose fp points to a file and has reached the end of the file. What is the return value of feof(fp) then? ( )
   A) EOF   B) −1   C) A nonzero value   D) NULL

6. [File buffer]

   The prototype of the function that reads a binary file is as follows: fread(buffer, size, count, fp); What does buffer refers to?
   A) Number of bytes in a memory block.
   B) An integer variable which represents the number of bytes in the data to be read
   C) A file pointer pointing to the file to be read
   D) The beginning address of a memory block, which represents the address of the data to be read

5.9.2 Fill in the tables

Fill in the tables in Figures 5.9 and 5.10 based on the following programs:

①scanf input	-1	0	1	2	3	a
①scanf return value						
①temp						/
②123.dat						/

Figure 5.9: Files: fill in the tables question 1.

①str value	"123abcDEF"
②test.txt content	
③strvalue	

Figure 5.10: Files: fill in the tables question 2.

1.   [fprintf and fscanf]
Suppose there is file "123.dat" under the same directory of the program below. The file is empty.

```c
#include <stdio.h>
int main(void)
{
  FILE * fp = NULL;
  int temp;
  fp = fopen("123.dat", "w");
  while(scanf("%d",&temp)) //——①
  {
    fprintf(fp, "%d", temp); //——②
  }
  fclose(fp);
  fp = fopen("123.dat", "r");
  while (fscanf(fp,"%d",&temp) != EOF)
  {
    printf("%5d", temp);
  }
  fclose(fp);
```

```
  return 0;
}
```

2. [fputc and fgets]

```
#include<stdio.h>
int main(void)
{
 FILE *fp;
 char str[100];
 int i=0;
 if((fp=fopen("test.txt", " w " ))= =NULL)
 {
  printf("Can't open this file.\n");
  exit(0);
 }
 printf("Input a string: \n");
 gets (str); //————①
 while (str[i])
 {
  if(str[i]>= 'a'&&str[i]<='z')
  str[i]= str[i]-32 ;
  fputc(str[i], fp); //————②
  i++;
 }
 fclose (fp);
 fp=fopen("test.txt", "r" );
 fgets(str, 100, fp); //————③
 printf("%s\n", str);fclose (fp);
 return 0;
}
```

Suppose that the keyboard input is "123abcDEF"

5.9.3 Programming exercises

1. Read/write formatted file
Suppose that contents of text file 20083.txt and text file 20084.txt are all real numbers in the format "xx.x." Please write a program that combines data in these files and save them to file 20085.txt (data in 20083.txt followed by data in 20084.txt).

2. Read/write data blocks in file
We have data of 10 students (including student IDs, names, and grades in three courses) in the file "score.txt." Please write a program that stores data of students

who failed at least one course into the file "bhg.txt." and data of students who passed all courses into the file "hg.txt."

3. Read characters from file

Please write a program that opens a specified file (if it exists), read data in blocks of 128 bytes, and prints each block on the screen in hexadecimal form and in ASCII values.

Appendix A   Adding multiple files to a project

There is more than one method to add multiple files to a project. We shall introduce one of them through the following example:

Example A C program consists of three files, namely testfile1.cpp, testfile2.cpp, and testfile3.cpp. We wish to add them to the same project.

testfile1.cpp:

```
/*The program consists of 3 files, main function is in test file 1*/
01 #include <stdio.h>
02 extern int reset(void);    /* Declare reset as an external function* /
03 extern int next(void);     /* Declare next as an external function */
04 extern int last(void);     /* Declare last as an external function */
05 extern int news(int i);    /* Declare news as an external function */
06
07 int i=1;                   /*Define global variable i*/
08 int main()
09 {
10    int i, j;               /*Define local variables i and j*/
11    i=reset();
12    for (j=1; j<4; j++)
13    {
14       printf("%d\t%d\t",i,j);
15       printf("%d\t",next());
16       printf("%d\t",last());
17       printf("%d\n",news(i+j));
18    }
19    return 0;
20 }
```

testfile2.cpp:

```
01 extern int i;             /*Declare global variable i*/
02
03 int next(void)
04 {
05   return ( i+=1);
06 }
07
08 int last(void)
09 {
10   return ( i+=1);
11 }
12
13 int news( int i)          /*Define parameter i, which is a local variable*/
```

https://doi.org/10.1515/9783110692303-006

```
14 {
15   static int j=5;          /*Define static variable j*/
16   return ( j+=i);
17 }
```

testfile3.cpp:

```
01 extern int i;              /*Declare global variable i*/
02 int reset(void)
03 {
04   return ( i );
05 }
```

Terms:
- Internal function: an internal function is only accessible by functions in the same file. We use keyword static to define internal functions. They are also called static functions.
- External function: if we define a function with keyword extern, then this function is an external function. For example:

```
extern int reset(void);
```

Other files can call the function reset. If we omit extern in a function definition, then the function is by default external.

   In the file where we make a call to an external function, we should use extern to indicate that the function is external.

Difference between declaration and definition: when a function or a variable is declared, no physical memory is allocated to it. Declarations make sure that our programs can be compiled. When a function or a variable is defined, it takes up physical space in the memory. A function or a variable can be declared multiple times. However, it can be defined only once.

   How do we access global variables when there are multiple files? If we define a global variable in one file and want to access it in other files, we must declare the global variable with keyword extern. If a global variable is defined with static, however, then it can only be accessed in the same file instead of by other files.

   To add multiple files to a project, we should follow these steps:

(1) Create a new project (suppose the name is test) in the IDE, as shown in Figure A.1

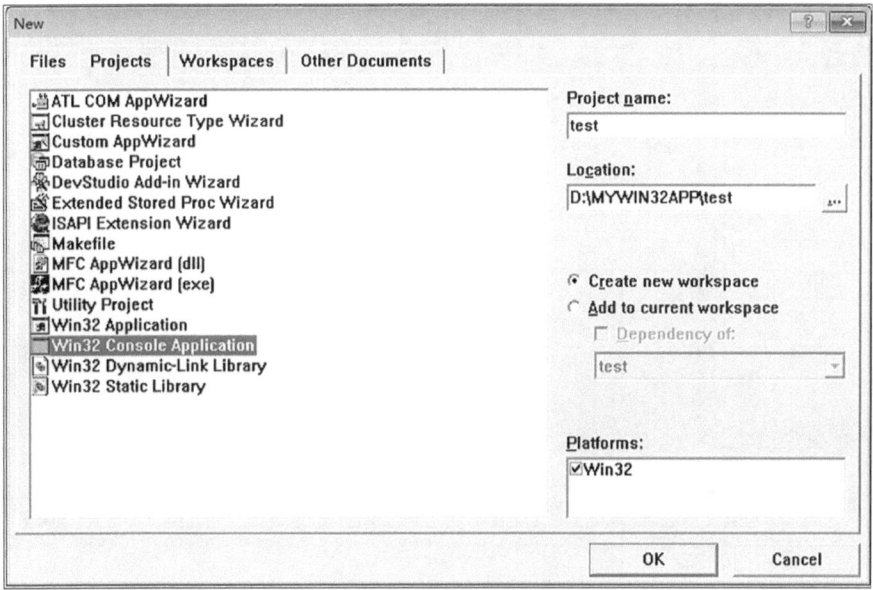

Figure A.1: Create a new project.

(2) Create a new file "testfile1.cpp" in the project test, as shown in Figure A.2.

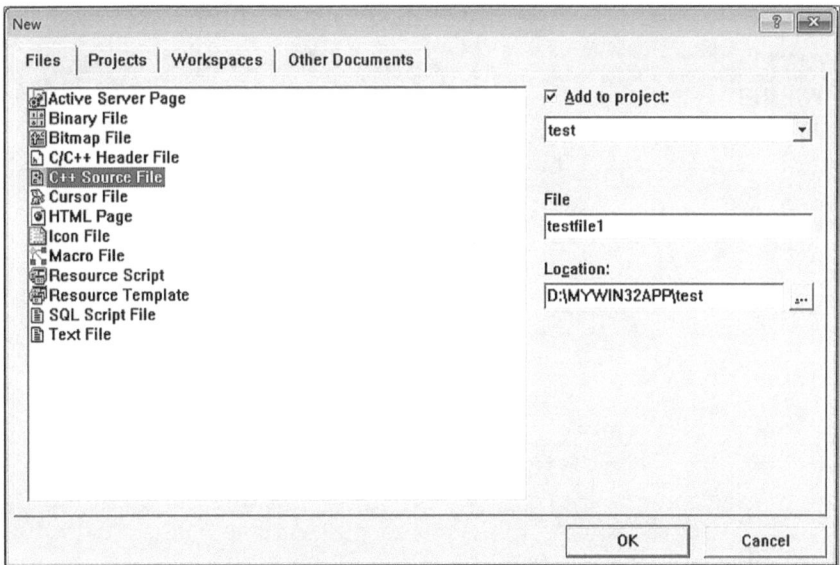

Figure A.2: Create a new file "testfile1.cpp".

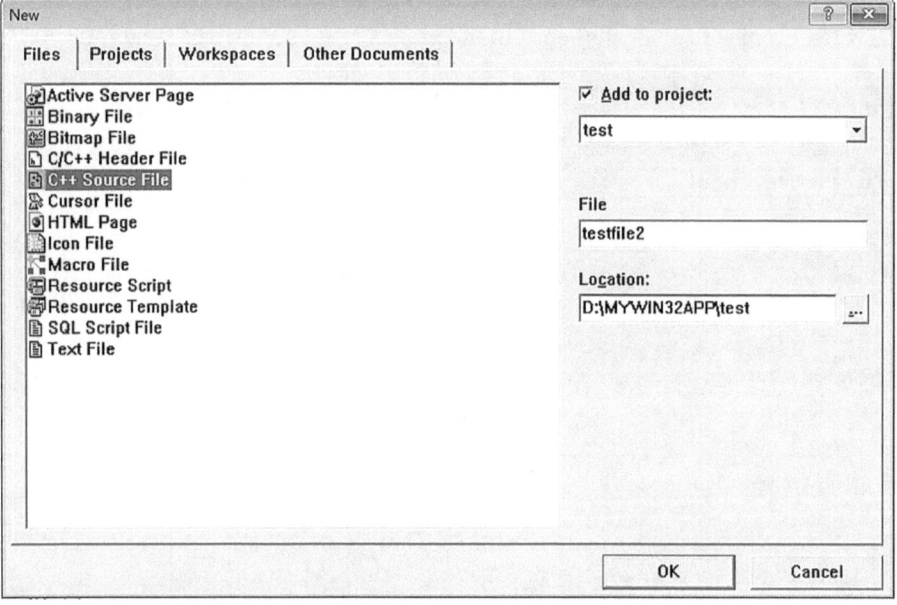

```
File Edit View Insert Project Build Tools Window Help

                                                    test          Win32 Debug

  test classes        testfile1.cpp *
                       /*The program consists of 3 files, main function is in test file 1*/
                       #include <stdio.h>
                       extern int reset(void);      /*Declare reset as an external function*/
                       extern int next(void);       /* Declare next as an external function */
                       extern int last(void);       /* Declare last as an external function */
                       extern int news(int i);      /* Declare news as an external function */

                       int i=1;                      /*Define global variable i*/
                       int main(void)
                       {
                           int i, j;                 /*Define local variables i and j*/
                           i=reset();
                           for (j=1; j<4; j++)
                           {
                               printf("%d\t%d\t", i, j);
                               printf("%d\t", next());
                               printf("%d\t", last());
                               printf("%d\n", news(i+j));
                           }
                           return 0;
                       }
  ClassView  FileView
```

Figure A.2 (continued)

(3) Create a new file "testfile2.cpp" in the project test, as shown in Figure A.3.

```
New

  Files │ Projects │ Workspaces │ Other Documents

   Active Server Page                        ☑ Add to project:
   Binary File
   Bitmap File                                test
   C/C++ Header File
   C++ Source File
   Cursor File                                File
   HTML Page                                  testfile2
   Icon File
   Macro File                                 Location:
   Resource Script
   Resource Template                          D:\MYWIN32APP\test
   SQL Script File
   Text File




                                              OK            Cancel
```

Figure A.3: Create a new file "testfile2.cpp".

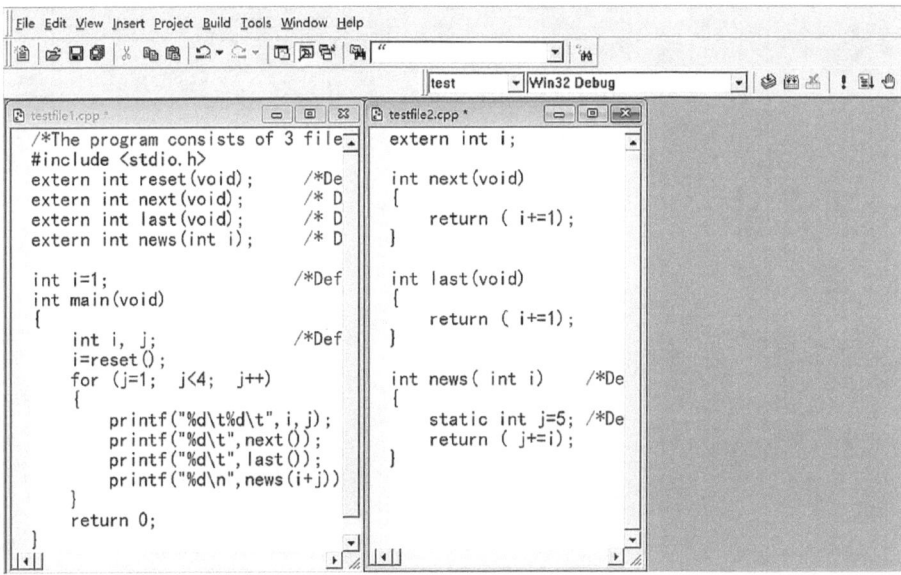

Figure A.3 (continued)

(4) Create a new file "testfile3.cpp" in the project test, as shown in Figure A.4.

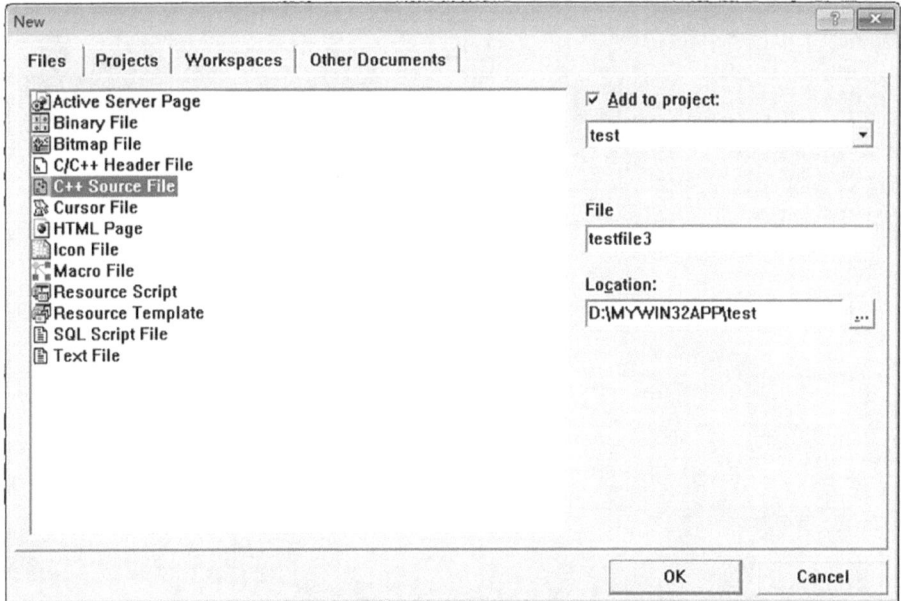

Figure A.4: Create a new file "testfile3.cpp".

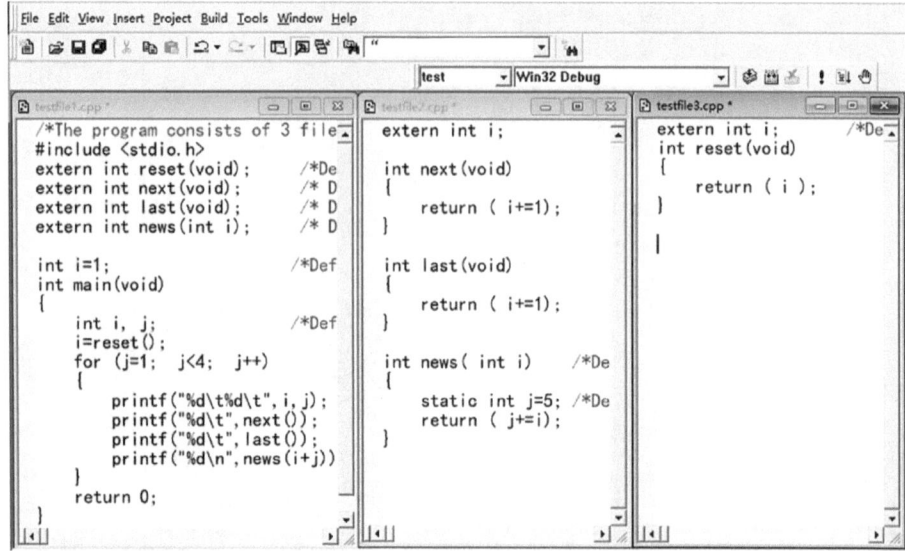

Figure A.4 (continued)

(5) Compile "testfile1.cpp," as shown in Figure A.5.

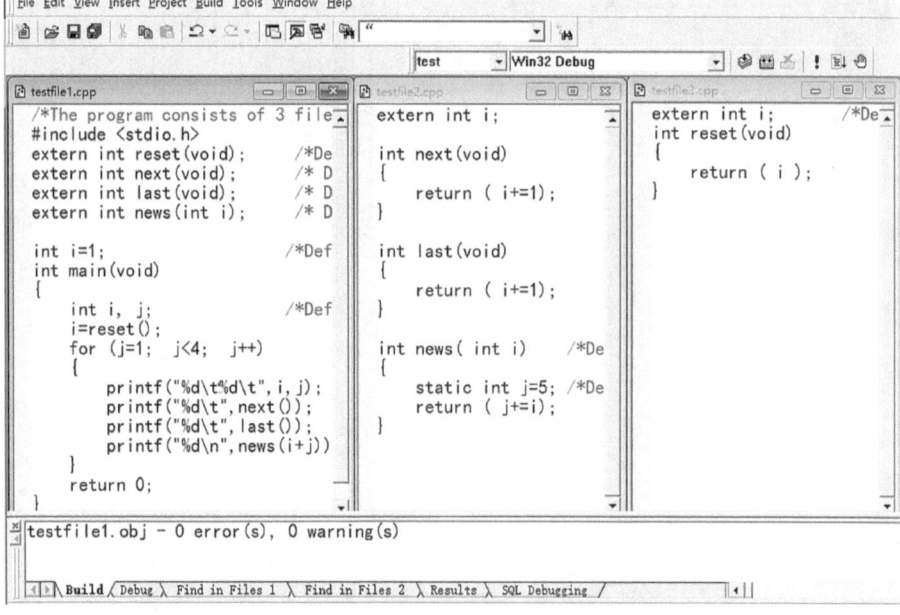

Figure A.5: Compile file "testfile1.cpp".

(6) Compile "testfile2.cpp," as shown in Figure A.6.

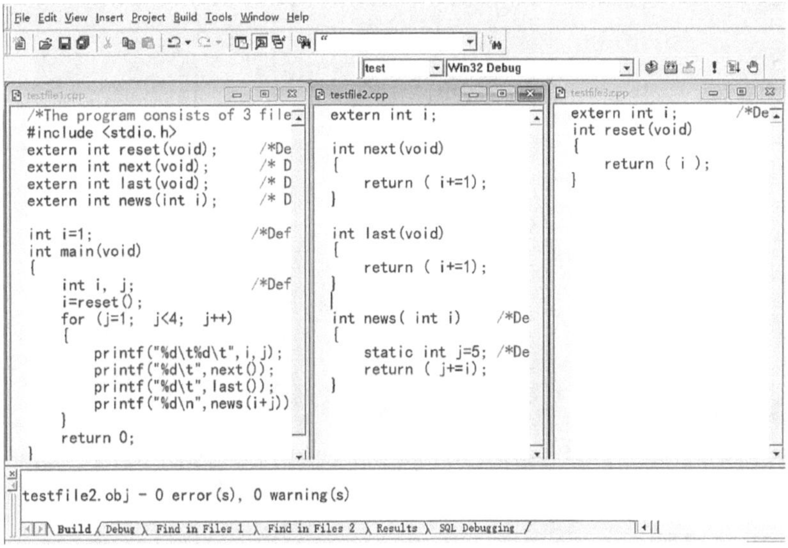

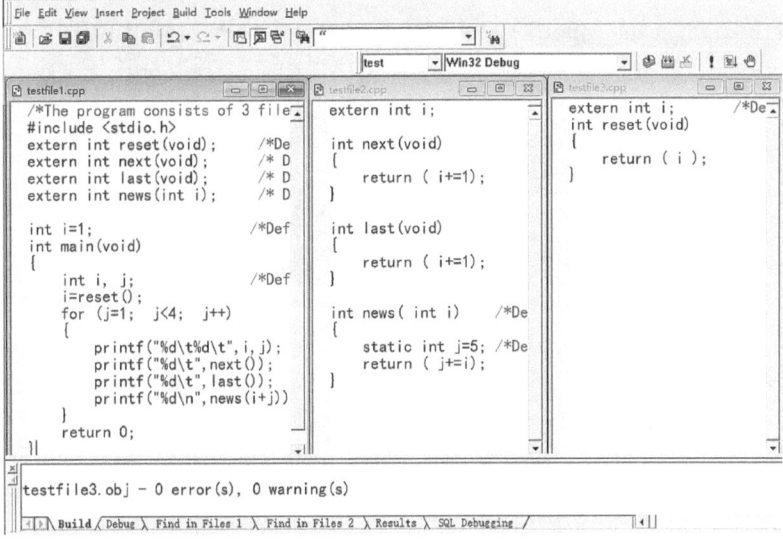

Figure A.6: Compile file "testfile2.cpp".

(7) Compile "testfile3.cpp," as shown in Figure A.7.

Figure A.7: Compile file "testfile3.cpp".

(8) Execute "Build" command in the window of the main function to generate an executable exe file, as shown in Figure A.8.

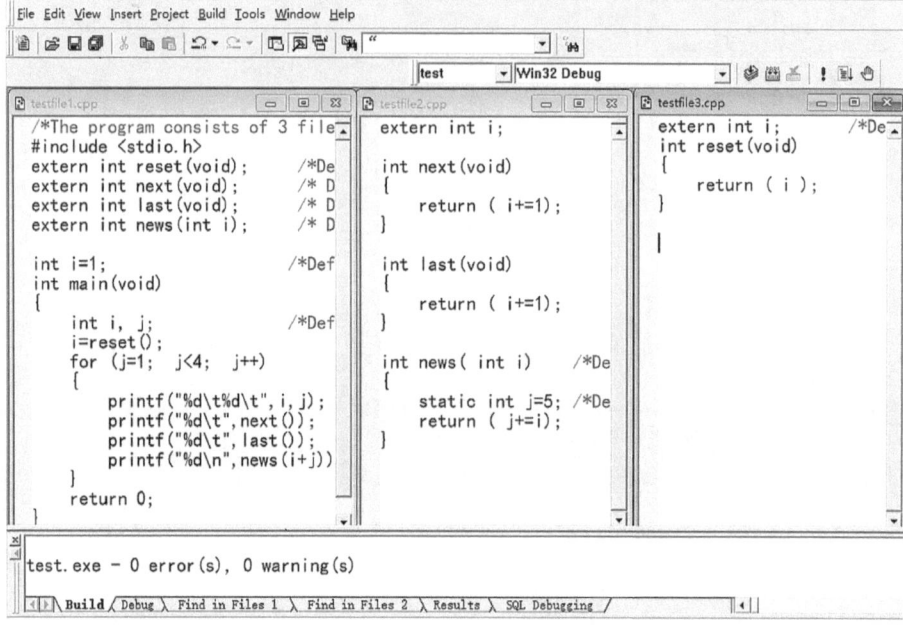

Figure A.8: Generate executable exe file.

(9) Run the program in the window of the main function to obtain result.

Program result:

```
1  1  2  3  7
1  2  4  5  10
1  3  6  7  14
```

Appendix B Programming paradigms

We write codes to solve practical problems. Programmers have different worldviews and methodologies when creating virtual worlds. There are multiple ways to solve problems. We call effective and universal patterns of problem-solving paradigms.

To be more specific, programming paradigms refer to styles and patterns of programming. Each paradigm guides us to analyze and solve problems in a unique way.

Paradigms exist in daily life, too. Formatted forms like deposit slips, withdrawal slips, receipts and invoices require us to fill in them in a specific way.

As shown in Figure B.1, programming paradigms include the imperative paradigm and declarative paradigm. We can classify languages that are not imperative into the declarative category. A paradigm may be used in many programming languages, and a programming language may support multiple paradigms as well. Fundamental programming paradigms include procedural, object oriented, functional, and logic.

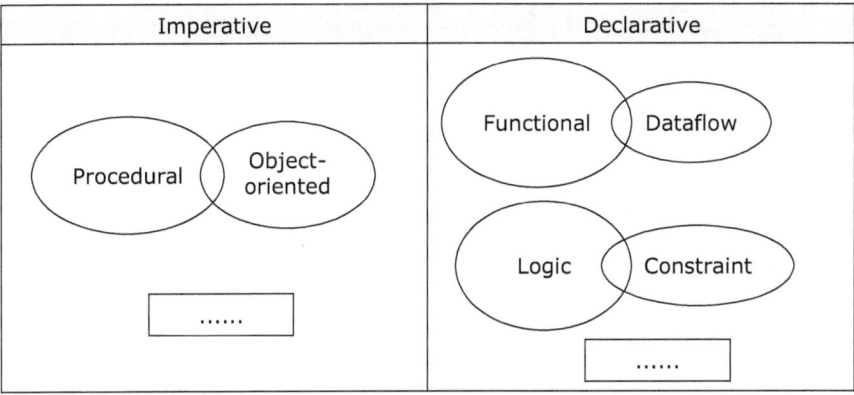

Figure B.1: Programming paradigms classification.

An imperative program is a sequence of commands of a Von Neumann machine. Object-oriented paradigm uses independent code blocks and objects and drives them through messages. Functional programming is an abstraction of mathematics, which describes computation as the evaluation of mathematical functions. Logic programming uses facts and rules to derive and prove conclusions.

The popularity of a programming language is closely related to its areas of expertise. For example, functional and logic languages are good at applications that use mathematics and logic, such as artificial intelligence, symbol handling, databases, and compilers. User-oriented applications are mostly interactive, event-driven, and have different business logic, so it is better to use imperative languages for them.

There is no absolute border between declarative languages and imperative languages. They are all constructed based on low-level languages. Thus, they can be used

https://doi.org/10.1515/9783110692303-007

together. For example, the introduction of functions or procedures in imperative languages makes them more declarative.

Fortran, COBOL, Pascal, and C are common procedural languages; Smalltalk, Java, C++, and C# are all object-oriented languages; Lisp, Haskell, and Clean are functional languages; logic languages include Prolog and others.

B.1 Procedural programming

A procedure is a set of steps (or commands) to complete a particular task.

Procedural programming uses a series of commands to implement these steps and complete the required job.

There are many "procedure-oriented" examples in daily life. For example, the process of "going to the outpatient department of the university hospital" consists of several steps, as shown in Figure B.2. Each step can be seen as an independent procedure. In other words, we do one thing at a time. Such a step is called a "module" in programming. (Note: suppose we pay in one go after we are done with the process.)

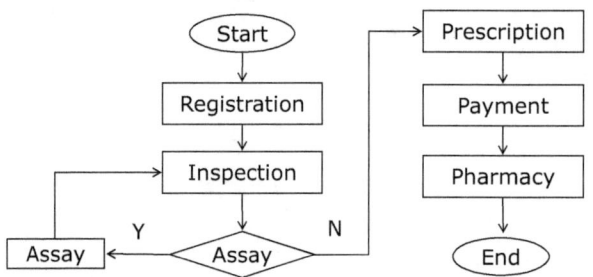

Figure B.2: The first process of "going to the outpatient department of the university hospital".

- Module: a module is a collection of statements that has its own name and can complete specific tasks independently. The internal implementation of a module is invisible from the outside. A module communicates with the outside through information interfaces.
- Information interface: an information interface describes how other modules and programs use this module. Information in an interface includes input/output information.

In the earlier example, "prescription" and "assay results" are the interface information. The prescription is used in several modules. We call data that are available to all modules "global variables."

Procedural languages are imperative languages with the addition of child programs. Because all modern imperative languages have this feature, we often use these two terms interchangeably. Terms like "procedure," "child program," "function," and "module" refer to the same thing in programming.

"Procedure-oriented" is a programming paradigm that focuses on modules. In procedure-oriented programming, we use a top-down stepwise refinement development method to divide a complex system into several independent child modules. Then we determine how these modules are assembled and how they interact with each other (that is, how they call each other). After designing these child modules, we combine them together to get the final system. Each module is implemented by fundamental structures like sequential, branch, and loop structure.

The procedure-oriented paradigm got its idea from sequences of computer commands. It converts solutions to problems into conceptualized steps and then translates these steps into program instruction sets, in which instructions are listed in a specific order.

A procedural program consists of three components, as shown in Figure B.3. The calling rules of modules describe the execution order of modules. In a single run of a program, the control is exchanged between the calling program and the program being called, as shown in Figure B.4.

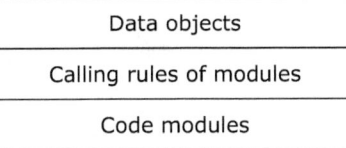

Figure B.3: Structure of a procedural program.

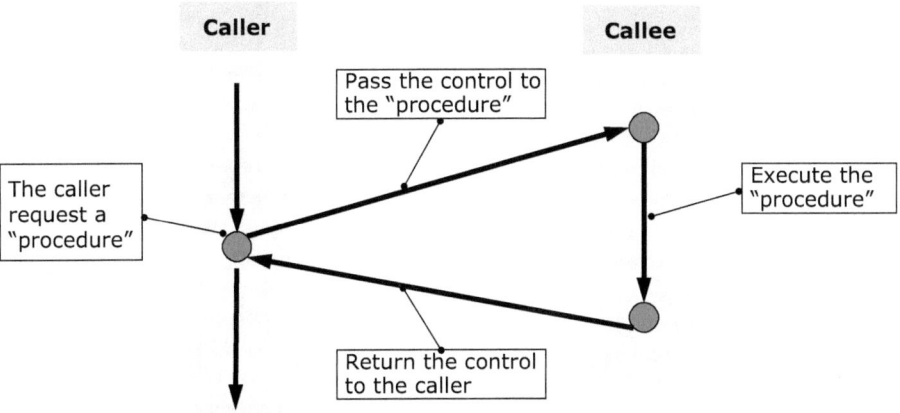

Figure B.4: Flow control of procedure calling.

A function call hands the execution of a program (usually a child program) over to other modules and preserves the context of the calling program. After the program being called terminates, it returns to the saved context.

The nature of procedure-oriented programming is dividing problems into modules. It solves practical problems based on the characteristics of problem-solving with computers. A multimodule system does module calling following a predetermined routine.

When the scale of such a system is large, it is difficult to modify it if we have different needs later. In the hospital example above, if we want to move "assay" after "payment," we have to update many components of the flow. If the system is extensive, it would be hard to maintain. Besides, if we put no restriction on global data, errors in them will affect other components of the system. "Prescription" is the global data in the hospital example. If the pharmacy does not have the medicine listed on the prescription, exceptions will occur in many modules. A summarization of critical issues in procedure-oriented programming is given in Figure B.5.

Nature	Function design	Designers' perspective: design programs based on the characteristics of computers
		Computation process: module call
Charac-terisTics	A modular structure that focuses on functions	Divide complex programs into simple and independent procedures
	Hard to maintain large systems	e.g., move "assay" after "payment"
	No restriction on accessing global data	Errors in global data affect many components of the system, e.g., medicine on the prescription is out of stock
	Execution pattern	Executed in a predetermined order

Figure B.5: Critical issues in procedure-oriented programming.

B.2 Object-oriented programming

Let us take a look at a story of humans and animals. There is a cage, as shown in Figure B.6. We can put at most one animal in the cage. If the cage is empty, the hunter may put a monkey into it and inform the zoo; the farmer may put a pig into it and inform the restaurant; the zoo would like to purchase the monkey, and the restaurant would like to buy the pig. Can we simulate this process using a program? We will find that we never know who is going to act first. In other words, we do not know the calling order of modules, which is necessary for procedure-oriented programming.

In the procedural paradigm, modules are executed in a predetermined order and in a flow-driven manner. The steps in the solution designed by programmers should be executed one after another. The system structure depends on our task. A change in one module may require changes in all related modules.

Figure B.6: A story between humans and animals.

In the example above, the execution order of modules depends on the state of the system. Execution of modules is thus driven by events or messages. "Fetching or putting animals" are events, and "informing" is a message. Figure B.7 shows an abstraction of this example. Procedural thinking does not work in such event-driven problems, so people have to look for other programming paradigms. Therefore, object-oriented programming comes to being.

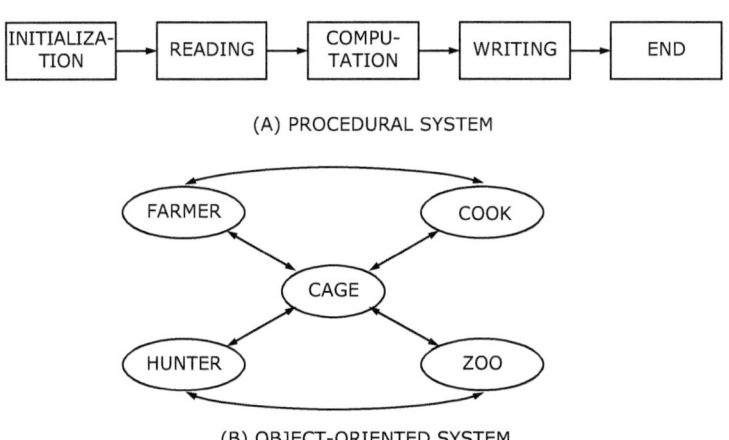

Figure B.7: Procedure oriented and object oriented.

If we use the event-driven method to describe the hospital example, we will find that the nature of modules in the system has changed. In Figure B.8, steps that were initially carried out by patients are now operations carried out after information exchange by departments of the hospital. The functionality extracted from the problem is no longer the core. Instead, the focus has become entities like departments. The events in this example are the reception of specific certificates, such as medicare card, registration form, and prescription. A patient must have a medicare card and money to register. A prescription cannot be filled without the assay sheet with payment proof. These certificates drive operations and information exchange among entities.

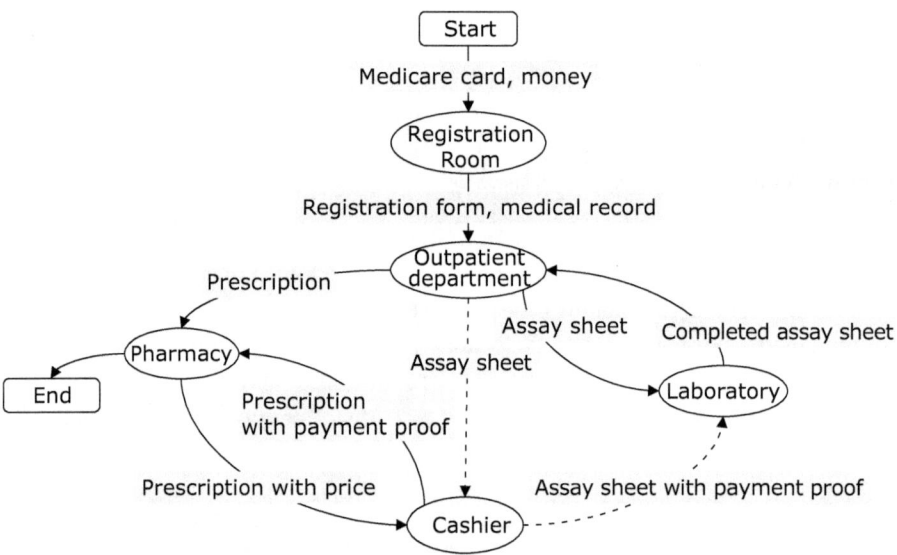

Figure B.8: The second process of "going to the outpatient department of the university hospital".

An entity has attributes (data) and behaviors (functionality). To make an object work, we need corresponding certificates. In object-oriented programming, these entities become objects. An object includes data and operations on its data. The certificates become messages (or events). Communication and control of objects are done by sending and receiving messages, as shown in Figure B.9.

An object-oriented system is built upon the interaction between objects. The execution of modules is driven by events or messages. The impact of changing a module is often limited to the module. In the hospital example, if we put "assay" after "payment," we only need to send the "assay sheet" message to the cashier instead of the laboratory, and then send the "assay sheet with payment proof" to the lab. Dashed lines in Figure B.8 represent this change.

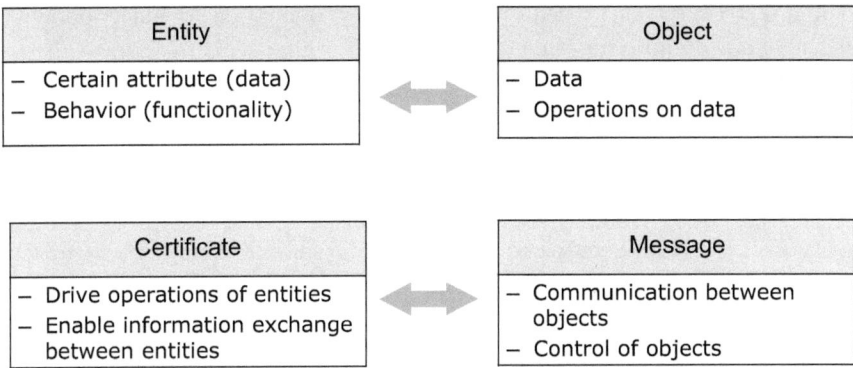

Figure B.9: Abstraction of object-oriented concepts in practice.

Entities in a hospital can work in parallel. However, modules are executed sequentially on computers with a single CPU. How do we execute these parallel operations on computers? The solution is to store messages in a queue and handle them sequentially, as shown in Figure B.10. An event generates a message, which is received by the system. The event scheduler fetches messages from the message queue and triggers corresponding procedures. Because the CPU works at high speed, it seems like the machine is doing multiple works simultaneously.

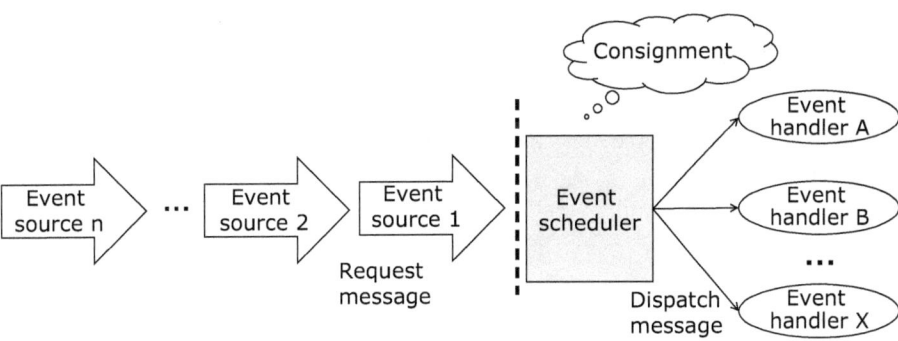

Figure B.10: Event-driven mechanism in computers.

In the Windows system, messages are the fundamental communication method. Events are major sources of messages. The occurrence of an event is known through messages. Whenever the user triggers an event, for example, a mouse movement or a keystroke, the system converts it to a message and stores it in the message queue of the corresponding program. The program fetches messages using GetMessage, preprocesses them using TranslateMessage, and eventually send them to the window process WndProc using DispatchMessage.

Figure B.11 compares procedure-oriented programming with object-oriented programming and summarizes the major differences.

Procedure-oriented methods	Object-oriented methods
(1) Focus on functionality	(1) Focus on objects
(2) Isolate data and processing routine	(2) Encapsulate data and operations in objects
(3) Re-develop similar software	(3) Inheritance of classes
(4) Execution order is predetermined	(4) Execution state of an object is controlled by message

Figure B.11: Comparison of procedure-oriented programming and object-oriented programming.

"Object-oriented" is an idea that focuses on entities. Data and operations on them are encapsulated into a single entity: an object. We summarize common features of similar objects in a class. Most data of a class must be processed using methods in the same class. Classes interact with the outside world through interfaces. Objects communicate with each other through messages. Users determine the execution order of programs.

The original intention of introducing "object-oriented" is to isolate "interface" from "implementation," so that lower level changes do not affect upper level functionality. Object-oriented languages describe systems more naturally. It is easier to reuse code and update our applications using these languages. Major features of object-oriented languages include:

(1) Identifiability: basic components in the system can be identified as discrete objects;

(2) Classifiability: objects with the same data structure and behaviors can form their own class;

(3) Polymorphism: an object has a unique static type and multiple possible dynamic types;

(4) Inheritance: data and operations can be shared among classes on different levels.

The first three are the fundamental features, while the last one makes object-oriented languages different from others. They (sometimes with dynamic binding as well) make a strong expression ability possible. The components in an object-oriented program are objects, each of which has its own attributes and behaviors. Computation is done by creating new objects and communication between objects.

B.3 Functional programming

Functional programming is a mathematical thought that describes computations as functions.

In functional programming, a program is a mathematical function, which is a black box that maps inputs to output, as shown in Figure B.12. We call it a black box because we do not know the implementation of it.

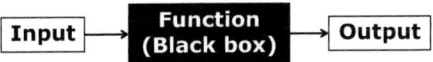

Figure B.12: Functions in functional programming.

The term "function" in functional programming is no longer a subroutine in computers. Instead, it refers to a mathematical function, namely a mapping of variables in its domain.

Functional programming languages have the following features:
- a functional language should define a series of primitive (atomic) functions for programmers to use;
- a functional language should allow programmers to build new functions using primitive functions.

The major use cases of functional languages are in mathematical derivation and parallel computing. They are suitable for solving mathematics problems of a limited scale.

Example of functional programming 1
Mathematical expression: (1+2)*3−4 may be translated into the following procedural language code:

```
var a = 1 + 2;
var b = a * 3;
var c = b - 4;
```

Functional languages require us to use functions. We can define different operations in the computation as functions, and rewrite the process as follows:

```
var result = subtract(multiply(add(1,2), 3), 4);
```

Example of functional programming 2
Scheme defines a series of primitive functions. The function name and input list are written in parentheses, and the result is a list that can be used as input of other functions.

For example, function car fetches the first element of a list. Function cdr fetches all elements from a list except the first one. They are used as follows:

```
(car 2 3 7 8 11 17 20)-> 2
(cdr 2 3 7 8 11 17 20)-> 3 7 8 11 17 20
```

We can combine them to obtain the third element of a list: (car (cdr (cdr List))). If the list is 2 3 7 8 11 17 20, then the result is 7.

B.4 Logic programming

Logic programming is a programming paradigm that instructs us to write code based on the logic process of derivation and computation in our brain:

$$Algorithm = Logic + Control$$

Logic programming describes facts and makes rules for them. The process of designing programs is constructing a proof. The facts refer to the relations between objects and attributes. The rules describe relations between facts. The execution of a program is the process of derivation based on the rules. Logic programming is completely different from other paradigms.

Logic programming uses logic as programming languages, and considers computation as a programming technique of controlled derivation. Users only need to write the logic of their programs, and the control part is left to the interpreter program in the system. In conclusion, the programming process can be represented by this formula: facts+rules = results.

In 1972, Alain Comerauer's group invented the first logic programming language, Prolog. Prolog is suitable for artificial intelligence programs, which include expert systems (a program that generates a suggestion or answer using a sophisticated model), natural language processing, theory proving (a program that generates new theories as an expansion of current ones), and some intelligence games. Prolog is usually used with some other languages in projects, where logic operations are done in Prolog and components like computation and user interfaces are implemented in other languages.

Example of logic programming
Facts and rules construct programs in Prolog. For example, two facts about humans are as follows:

```
human (John)
mortal (human)
```

The user may check:

```
? -mortal (John)
```

The program will respond with yes.

Now we may make a summary of programming paradigms in Figure B.13.

Paradigm	Program	Input	Output	Programming object	Execution object
Imperative	Automata	Initial state	Final state	Statement	Command
Functional	Mathematical function	Independent variable	Dependent variable	Function	Expression evaluation
Logic	Logic proofs	Fact	Conclusion	Proposition	Logic derivation

Figure B.13: Summary of programming paradigms.

Appendix C   void type

There are many uses of "void type" in C. In different cases, void type has different meanings.

1. void type
The type specifier is void. void type does not refer to a specific data type. Instead, it is used when a function has no return value or to represent a generic pointer.

2. void-type functions
When a function is called, it usually returns a value to its caller. The value must have a data type and should be specified in the function definition and declaration. However, some functions do not return any value to their callers. These functions can be defined as "void type."

3. void-type pointers and null pointers
(1) void-type pointers
A void pointer is also called a generic pointer or a pointer with no type associated with it. The data stored in the memory block pointed by a void pointer can be of any type available in C.

Why do we need void pointers? Because sometimes we do not know what data we are going to store in the memory block pointed by the void pointer. Thus, a unique mechanism is needed. For example, malloc requests a continuous memory block dynamically during program execution. The return value of malloc is a pointer to this block. The designer of malloc has no idea what users will store in the allocated memory, so he/she has to make the return value a void pointer to accommodate all cases.

Some languages have a dedicated type for pointers. The merit of doing so is that it does not matter what data will be stored in the memory block allocated.

We cannot use void pointers to access data unless we convert it to a typed pointer forcibly.

(2) Null pointer
Note that a null pointer is not a void pointer. A null pointer has value NULL. It does not point to any memory address. In the malloc example, if memory allocation fails, NULL will be returned as an indicator of exception.

https://doi.org/10.1515/9783110692303-008

Index

https://doi.org/10.1515/9783110692303-009